Sweet Sorghum

Production, Processing and Utilization

Sweet Sorghum

Production, Processing and Utilization

U.D. Chavan

Senior Cereal Food Technologist
Sorghum Improvement Project,
Mahatma Phule Krishi Vidyapeeth, Rahuri-413 722
Phone: 02426 243670 (R), 02426 243253 (O)
E-mail: uttamch2002@yahoo.co.in

J.V. Patil

Senior Sorghum Breeder
Sorghum Improvement Project,
Mahatma Phule Krishi Vidyapeeth, Rahuri-413 722
Phone: 02426 243605 (R), 02426 243249 (O)
E-mail: patiljvp@yahoo.co.in

ibdc publishers

Published by

ibdc publishers

Khushnuma Complex Basement
7, Meerabai Marg (Behind Jawahar Bhawan),
Lucknow 226 001 U.P. (INDIA)
Tel. : 91-522-2209542, 2209543, 2209544, 2209545
Fax : 0522-4045308
E-Mail : ibdco@airtelmail.in

First Edition 2010

Price: 750

ISBN 978-81-8189-470-0

Composed & Designed at :

Panacea Computers
3rd Floor, Agarwal Sabha Bhawan, Subhash Mohal
Sadar Cantt. Lucknow-226 002
Phone : 0522-2483312, 9335927082, 9452295008
E-mail . prasgupt@rediffmail.com

Printed at:

Salasar Imaging Systems
C-7/5, Lawrence Road Industrial Area
Delhi - 110 035
Tel. : 011-27185653, 9810064311

This scientific book is dedicated to our knowledgeable Gurus who taught us a, b, c in the scientific world.

U.D. Chavan
J.V. Patil

Foreword

Sorghum is mostly cultivated in *kharif* and *rabi* seasons in India. Maharashtra ranks first in the production of sorghum grain crop. The production of grain sorghum is higher per hectare in *kharif* season than the *rabi* season, because during *kharif* season, due to occurrence of rain, the water availability is more as well as the climate is warm and humid, which is inducive for sorghum growth. Similarly, Agricultural Universities also developed several hybrid genotypes, which yield very high-level grain production.

The *kharif* season sorghum is often vulnerable to rain, gets deteriorated due to mould attack because of the rains mostly occurring during *kharif* season and it makes sorghum grains unfit for human consumption. This ultimately results into the economic loss to farmers.

Such type of problems can be overcome by cultivation of sweet sorghum in *kharif* season. Sweet sorghum is capable of producing high biomass 70-80 per cent of which is the stalk having high content of juice (40-50%). The sweet sorghum gives a good quality fodder, syrup, jaggery, glucose/fructose syrup as well as ethanol as a biofuel. These products give higher economical returns to farmers. The authors Dr. U.D. Chavan and Dr. J.V. Patil have given an up-to-date and scientific information regarding sweet sorghum cultivation, production and utilization in the book on **"Sweet Sorghum (Production, Processing and Utilization)"**. I heartily wish them best of luck for publication of a book on "sweet sorghum", which will be beneficial to farmers, industrialists as well as students and scientists in the field of agriculture.

(Dr. R.B. Deshmukh)
Vice-Chancellor
M.P.K.V., Rahuri

Preface

During last decade the *kharif* season sorghum is often vulnerable to rains gets deterioration due to mold attack and becoming red or black grains because of the occurrence of rains during *kharif* season. Due to the rainy season, water accumulation occur in the glumes, which dissolve the pigmented material and spread over all grains. Due to moist condition, grains attacked by various types of fungus growth. Therefore, such type of sorghum grains become unfit for human consumption. This results into the loss of income from sorghum to the farmers.

Such type of problems can be overcome by developing new genotypes, which will resistant to grain mold. Sweet sorghum cultivation in *kharif* season is another solution over grain mold susceptible cultivator of sorghum. Sweet sorghum is capable of producing high biomass, good quality fodder, stalk can be utilize for the production of juice that can be utilize for the preparation syrup, jaggery, fructose/glucose syrup or bioethanol. These value added products of sweet sorghum gives higher returns to the farmers. The information regarding production, processing and utilization of sweet sorghum is not available in the form of book. Therefore, we tried to put up-to-date scientific information on **"sweet sorghum (Production, Processing and utilization)** in this book. We hope that the information given in this book will be helpful to the farmers, industrialists, scientists as well as the students in the field of agriculture.

U.D. Chavan
J.V. Patil

Acknowledgements

Authors are thankful to the Hon. Dr. R.B. Deshmukh, Vice-Chancellor, Mahatma Phule Krishi Vidyapeeth, Rahuri for his critical and fruitful suggestions while the preparation of **"Sweet Sorghum" (Production, Processing and Utilization)** book. We are thankful to those scientists who worked on sweet sorghum and their contribution is enclosed in this book as well as their sources are cited in the reference section. Authors are also thankful to all scientists and friends those helped directly or indirectly while preparation of this manuscript on sweet sorghum. We would like to express over profound thanks to our family members for their constant support, patience and devotion, which inspired us to complete this manuscript.

U.D. Chavan
J.V. Patil

About the Authors

Dr. U. D. Chavan obtained his B. Sc. and M. Sc. (Agri. in Biochemistry) degrees from Mahatma Phule Krishi Vidyapeeth Rahuri in 1985 and 1987 respectively. He received his Ph.D. degree in Food Science from Memorial University of Newfoundland St. John's Canada in 1999. He has done International training on "Global Nutrition 2002" at Uppsala University Uppsala, Sweden in 2002. He also attended follow-up International workshop on "Global Nutrition 2002" at Hanoi, Vietnam in 2002. Dr. Chavan visited Denmark, Finland, Ireland, France, Switzerland, Poland, Spain, Vietnam, Thailand, England, and U.S.A. under "Global Nutrition 2002" Programme sponsored by Swedish International Development Agency (SIDA). During this programme, he worked on human nutritional diet and disorders as well as on genetically modified organisms (GMO).

Dr. Chavan worked as Senior Research Assistant in the Department of Biochemistry and Food Science and Technology at Mahatma Phule Krishi Vidyapeeth Rahuri from 1988 to 2000. During his Ph. D., he worked as Technician/Research Associate at Atlantic Cool Climate Crop Research Center and Agriculture and Agri-Food Canada. He has guided 15 students for M.Sc. (Agri.) in Biochemistry and Food Science and Technology.

Dr. Chavan received Korgavkar Trust Fellowship, under-graduate and post-graduate merit scholarship as well as Senior Research Fellowship from ICAR, New Delhi and United States Department of Agriculture U.S.A. He received University Graduate Fellowship from Memorial University of Newfoundland St. John's Canada for his Ph.D. programme. Dr. Chavan also received international Scholar Award and Excellentiam Award for his Ph. D. research work from Memorial University of Newfoundland St. John's Canada. He has received a "Certificate of Appreciation" from the U.S. Department of Agriculture in 1992 for his work on processing of groundnut under the guidance of Dr. S. S. Kadam (Principle Investigator).

He has written 85 research papers and 40 popular articles. He has authored nine books in Marathi, three books in English and two book chapters in English. He has been awarded "Literary Award" for his best book on "Growth Regular" on agriculture. Dr. Chavan was selected as a best group leader and best presentation for "Global Nutrition 2002" by SIDA. He has been awarded "Life Time Achievement Award 2004" for his outstanding contribution in post-harvest technology of fruits and vegetables and allied fields by United Writers' Association of India. He has been elected as a Fellow of United Writers' Association of India in 2004. Dr. Chavan received Pride of Nation Award 2006 and Maharashtra Gunigan Ratna Award for 2006. He also received Jewels of India Award for 2006 for his contribution in the field of Food Science and Technology.

He has given 5 recommendations in the field of Food Processing. He also actively involved in the developement of NIAW-917 (Wheat), SPV-1546, RSV-1625 (Chitra), RSV-423 (Vasudha) and RSV-458 (Anuradha) Sorghum varieties. He has given 15 radio talks and 3 TV programmes. Currently he is working as Senior Ceral Food Technologist at Sorghum Improvement Project, MPKV, Rahuri.

Dr. J. V. Patil obtained his B. Sc., M. Sc. and Ph. D. degrees (Agri. in Botany, Cytogenetics and plant breeding) from Mahatma Phule Krishi Vidyapeeth Rahuri in 1979, 1981 and 1992 respectively.

Dr. Patil worked on several positions since 1981 as Senior Research Assistant, Jr. Genetist-2, Associate Professor, and Plant breeder, Professor of Botany and now from 2007, he is working as a Head Department of Agricultural Botany and holding additional charge of Sorghum Breeder at Mahatma Phule Krishi Vidyapeeth Rahuri. During his service carrier, he worked on several administrative posts in the University. He has guided 12 students for M. Sc. (Agri.) and 5 students for Ph. D. in the discipline of Agricultural Botany.

Dr. Patil received Merit Scholarships during his under and postgraduate study programme. He also received **Baliraja**: Late Annasaheb Shinde Smruti Krishi Sanshodhak Gaurao Puraskar (2004) and **"Vasantrao Naik Krishi Puraskar" (2005)** from Government of Maharashtra. He has written 73 research papers, 48 technical publications and 58 popular articles. He also organized 13 farmers' rallies, and given 27 radio talks and 10 TV programmes. He has developed Girna (safflower), Virat, Vaibhav, Varun, Vipula, Digvijay, Rajas (chickpea), Phule Amruta, Uttara, CSV 19 SS, Chitra, CMS 1409 A/B, RSV 423, (sorghum) crop varieties during his service carrier. He also gave 5 recommendations for farming various crops on farmers field. He also developed several innovative agro techniques such as planting methods, and sprinkler method in irrigated chickpea, technique for in-situ rainwater conservation in *Rabi* sorghum etc.

Abbreviations used in this book

Abbreviation	Full form
AICRP	All India Co-ordinated Research Project
AICSIP	All India Coordinated Sorghum Improvement Project
CBD	Convention of Biological Diversity
CCS	Commercial Cane Sugar
CSIR	Council of Scientific and Industrial Research
CSV	Coordinated Sorghum Variety
DUS	Distinctness Uniformity and Stability
GATT	General Agreement on Tariffs
HFS	High Fructose Syrup
HFSS	High Fructose Sorghum Syrup
IARI	Indian Agricultural Research Institute
ICAR	Indian Council of Agricultural Research
ICRISAT	International Crop Research Institute for South Asian Tropics
IPR	Intellectual Property Rights
IS	Indigenous Sorghum
M 35-1	Maldhandi
NAIP	National Agricultural Innovative Project
NARS	National Agricultural Research System
NATP	National Agricultural Technology Project
NBPGR	National Bureau of Plant Genetic Resources
NISCAIR	National Institute of Science Communication and Information Resources (http://www.niscair.res.in)
NRCS	National Research Council for Sorghum
NRS	Non-Reducing Sugars
PBR	Plant Breeder Rights
PVP	Plant Variety Protection
RS	Reducing Sugars
RSCR	Rahuri Sorghum Charcoal Rot
RSLG	Rahuri Sorghum Local Germplasm
RSP	Rahuri Sorghum Physiology

RSV	Rahuri Sorghum Variety
Sel-3	Selection 3
SPV	Sorghum Project Variety
TRIPS	Trade Related Aspects of Intellectual Property Rights
TS	Total Sugars
TSI	Total Sugar Index
WTO	World Trade Organization

Contents

Chapter 1

Introduction

The post rainy season (*Rabi* season) sorghum grain is mostly of good quality and used for human consumption. The *kharif* season sorghum is often vulnerable to rain gets deterioration due to mold attack because of the rains being occurred mostly during September-October period and it makes Sorghum grains unfit for human consumption. The utilization of such damaged sorghum grain in different form depends upon the percentage of blackening of the grain. If the percentage is less than 50% then it can be directly use for polishing or pearling and if the blackening percentage is more, then it must be treated with different chemicals to remove/reduce the mold from surface of the sorghum grain and then polish it with the help of rice polisher and then utilize for different purposes. Such type of processed rain damaged sorghum grain has enormous demand for industrial uses which include poultry or animal feed as well as production of starch, alcohol, dextrin's, glucose, high fructose syrup, malt etc.

Such type of problems can be overcome by developing new genotypes which will be resistant to grain mold and withstand in good condition in rainy season (*kharif* season). Sweet sorghum cultivation in *kharif* season is another solution over grain mold susceptible cultivars of sorghum. Sweet sorghum is capable of producing high biomass 70-80 per cent of which is the stalk having high content of juice (40-50 per cent). Sweet sorghum juice containing mostly sucrose, glucose, and fructose sugars. Such type of high sugary crop can be grown to produce the fuel alcohol. Currently, sugarcane industries are using molasses for the production of fuel alcohol by utilizing distilleries attached to sugar factories and created a major demand of alternate source. But recently there is a drastic reduction in the area of sugarcane, as water availability for growing sugarcane became a major constraint. Sweet sorghum can be cultivated in nearly all-temperate and tropical climatic areas. In addition to its wide range of adaptability, the crop is more profitable due to its several advantages.

1. Sweet sorghum has higher photosynthetic rate and biomass production capability than other cereal crops.
2. This crop required short duration, therefore it can be taken 2 to 3 times in a year.
3. Sweet sorghum crop can be sown/cultivated on any arable/cultivable land.
4. This crop required very low fertilizer doses as well as other input requirements.
5. Sweet sorghum crop is tolerant to biotic and abiotic factors such as draught, temperature, saline and alkali soil conditions etc.
6. Sweet sorghum stalk contain high juice percentage (40-50%) than other cereal crops and that can be processed for the production of fuel ethanol/alcohol.
7. The fermentable sugar is at higher level in sweet sorghum.
8. For ethanol production from sweet sorghumsame machinery can be used which is being used for alcohol production in the commercial sugar factories.
9. There will not be need to train new staff for ethanol production from sweet sorghum. Same staff from sugar factories can be used continuously.
10. Sweet sorghum bagasse has more biological value than the sugarcane bagasse; therefore it can be used as a fodder to the animals.
11. Ethanol produced from sweet sorghum or grains gives less pollution due to its less content of CO and octane number and it can mix very easily with petrol for combustion.
12. The mandated increase in ethanol blending levels from 5 to 10 per cent in second phase of Central Governments plan would trigger the manifold use of alternate raw ma-

terials. Since, there will be deficit of existing raw material viz., molasses. In such an event, it is important to make potential raw materials available through extensive research and extension efforts for popularization of sweet sorghum cultivation at various places for its suitability and feasibility.

Chapter **2**

Desired Characteristics of Sweet Sorghum Cultivars

For the maximization of Syrup/Jaggery/Sugar/Alcohol yield per hectare and to commercialize the cultivation of sweet sorghum with higher returns to the farmers; it is desirable to have following characters in sweet sorghum cultivars:

1. Ability to produce high yields of medium to large stalks per hectare (high biomass).
2. Few nodes and long internodes.
3. High percentage of extractable juice.
4. Excellent juice quality i.e. high percentage of fermentable sugars and low in starch, fiber, aconitic acid and colouring matters.
5. Adaptability to diverse soil and climatic conditions.
6. Resistant to lodging i.e. strong stalk, root system, and strong errect growth.
7. Tolerant to drought and fog.
8. Shorter growing period.
9. Resistant to diseases and pests.
10. Very low requirement of fertilizers and manures.
11. Possibility of crop harvesting by mechanization.
12. High accumulation rate of biomass within shorter period due to adopting C_4 plant photosystem.
13. Staling effect should be lower rate after harvesting.

Chapter 3

Comparative Characteristics of Sweet Sorghum and Sugarcane

Comparative characteristics of sweet sorghum and sugarcane crop are presented in Table 1 and Transformation of sweet sorghum into value added products is shown in Figure 1a and 1b.

Table 1. Comparative Points Between Sweet Sorghum and Sugarcane

Sr. No.	Sweet sorghum	Sugarcane
1.	Sweet sorghum can grow on rain fed area.	Sugarcane required irrigated heavy soil for cultivation.
2.	Sweet sorghum crop can grow twice a year.	Only one crop can grow in a year.
3.	Very less amount of water if available is need for sweet sorghum cultivation.	Large amount of water is required to sugarcane crop as compared to sweet sorghum.
4.	Sweet sorghum crop can grow on only organic fertilizer.	Sugarcane crop need both organic as well as inorganic fertilizers to get good yield.
5.	Small/marginal farmers can grow sweet sorghum due to less expenditure on cultivation.	Sugarcane crop need more expenditure, therefore small farmers needs to take loans from various sources.
6.	Sweet sorghum can be utilized for the production of fodder, syrup, gul or ethanol.	Sugarcane mostly utilized for the production of sugar and gul.
7.	Farmers can produce gul at small-scale level from sweet sorghum.	If sugarcane factories closed then farmers comes into very big difficulties.

8.	By cultivation of sweet sorghum farmers can gives employments to the large number unemployed youths.	Sugarcane industry gives employment to a small number of unemployed youths.
9.	Sweet sorghum can be utilized for the production of Murghas on large scale that can be utilized throughout the year.	Sugarcane cannot be utilized to produce Murghas on large scale.
10.	When sweet sorghum used as a fodder to milch animals, it increases milk production.	Sugarcane tops are not able to increase milk production.
11.	Seed rate and their cost is less than sugarcane.	Seed rate and their cost is very high than sweet sorghum.
12.	From dry sweet sorghum we can get both grains and straw.	Very small amount of dry sugarcane straw can be obtained from sugarcane and animals do not efficiently utilize that as a fodder.
13.	While preparation of various products from sweet sorghum there will be very less pollution.	High pollution in the sugar factory area and that affects human life.
14.	If farmers take 2 to 3 crops in years, it gives higher return than sugarcane.	Sugarcane gives less return because it takes 12 to 16 months to harvest for sugar factory.
15.	Sweet sorghum cultivated land need very less efforts to make it suitable for next crop.	Sugarcane cultivated land need heavy efforts and higher cost to make it suitable for the next crop.
16.	Grains and fodder obtained from sweet sorghum can be stored for next year.	Sugarcane cannot be stored for the next year.
17.	The leaves of sweet sorghum are not having hairy portion as well as no sharpness on edges so animals can eat very efficiently.	Sugarcane leaves have hairy portion and sharpness at the edges so animals hagitates to eat as a good fodder.
18.	Sweet sorghum when used as a fodder gives very high nutritive value for the nourishment of animal.	Sugarcane tops do not have high nutritive value so it is a poor quality fodder for animal.

19.	Expenditure on fertilizer is less than the sugarcane.	Very high expenditure on fertilizer than the sweet sorghum.
20.	Harvesting of sweet sorghum must be done at appropriate time only; otherwise there will be drastic reduction in juice yield. Staling effect is higher.	Such type of drastic reduction in juice yield will not occur in sugarcane. Staling effect is slower.
21.	Sweet sorghum canes must be crushed for juice extraction immediate after harvesting otherwise cane will dry quickly.	Such type of problem is not occurring in the case of sugarcane because canes are very sturdy than sweet sorghum.
22.	Sweet sorghum juice contains phenolics, chlorophyll, and other pigmented material so it gives colour to the syrup, jaggery products.	Sugarcane juice contains very less amount of such type of compounds so less problem than sweet sorghum.
23.	Sweet sorghum contain higher amount of proteins, starch and fibrous material in the juice that creates problem while jagger preparation.	Such type of problem is not occurring in the case of sugarcane juice.
24.	Sweet sorghum syrup and jaggery gives typical smell and taste.	Such type of characteristics is not observed in case of sugarcane syrup and jaggery.
25.	Sweet sorghum gives dark colour syrup and jagger than the sugarcane syrup and jaggery.	Sugarcane gives clear and transport syrup and pale yellow jaggery, which is more attractive than the sweet sorghum syrup and jaggery.

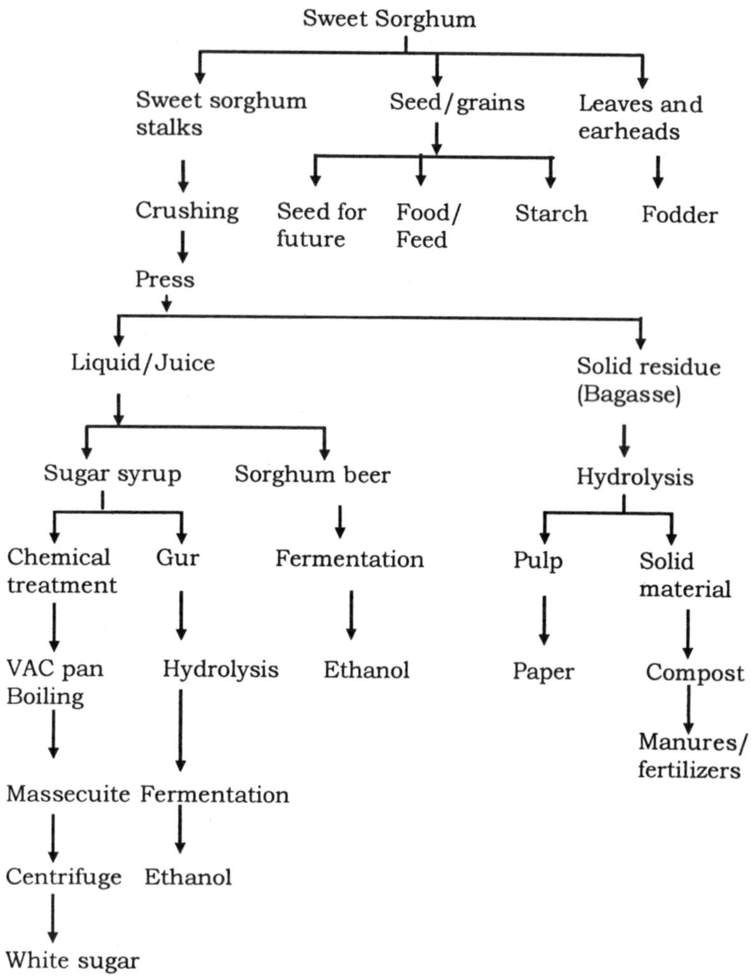

Fig 1a. Transformation of sweet sorghum into value added products

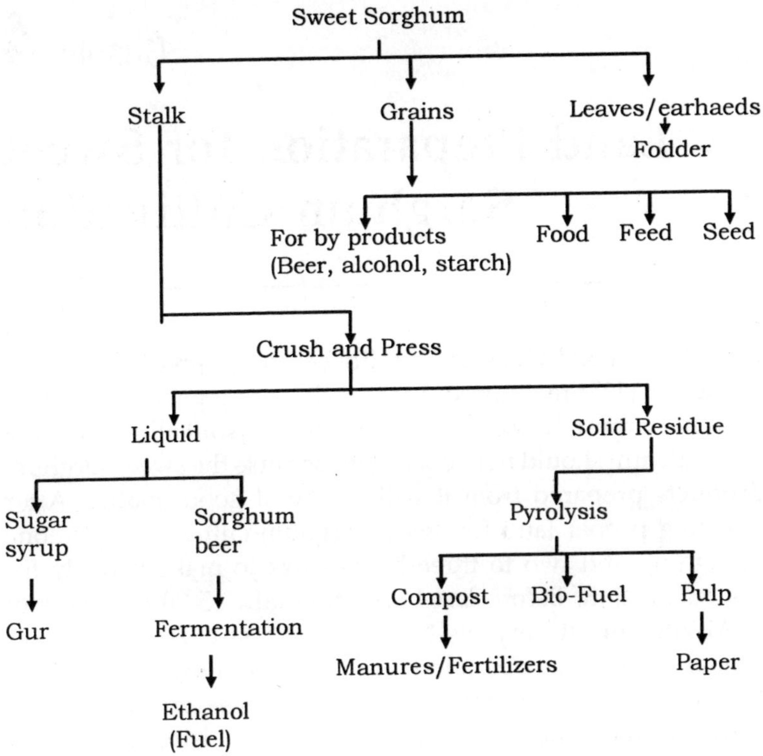

Fig 1b. Transformation of sweet sorghum into value added products

Chapter 4

Land Preparation for Sweet Sorghum Cultivation

For getting good yield and quality of grains, green cane yield or fodder yield medium to heavy soil having good drainage is most important. The soil containing high salts, alkaline or having high calcium should not be selected, because the sweet sorghum products prepared from it will not be of good quality. After selecting proper land for sweet sorghum cultivation give one ploughing and two to three harrowings to make it ready for the cultivation. Before last harrowing add 15-20 cartloads of FYM and mix it very well during the last harrowing. After preparatory tillage operation, prepare small boxes on the land before sowing in dryland area to conserve soil moisture as shown in the Plate 1. This conserved soil moisture is used for the good growth of the sweet sorghum crop.

Chapter **5**

Time for Sowing

The sowing of sweet sorghum should be done in last week of June or after commencement of monsoon. If water supply is available, then first irrigate the land and then do the sowing of sweet sorghum in the last week of June. For summer season, sowing should be undertaken in the last week of January. If water supply to the sweet sorghum is available in the summer season, then the second sowing can be done in the second fortnight of March. So that green fodder to the milch animals can be obtained in the month of May. If farmers want to take two crops in a year, then the sowing of sweet sorghum in *Rabi* season should be done in the month of September-October, then it will not required more water supply. The sowing of sweet sorghum should be done with the help of seed drill and the distance between two rows should be 30cm. The plant population must be maintained up to 140 to 160 thousands per hectare.

1. Seed Rate

Seeds of sweet sorghum must be sound having high percentage of germination. Before sowing, seed should be treated with sulphur. The seed rate 35 to 40 kg per hectare should be used for seed drilling sowing with 30cm row spacing.

2. Fertilizers Management

For higher yield and good quality of sweet sorghum grains, fodder as well as juice are mostly depends upon the good growth of plant and it mostly depends upon proper fertilizer application and water management as well as other agronomical practices. For one hectare 15 to 20 cartloads of cow dung or FYM should be applied at the time of last harrowing while land preparation. Application of 100:50 N:P kg/ha is recommended for sweet sorghum cultivation on medium to heavy soil. Half quantity of nitrogen and full quantity of phosphorus should be applied at or before sowing. Remaining half quantity of nitrogen is top-dressed at 25-30 days after sowing. If we apply more nitrogen than the recommended then the quality of juice and their products become disturb.

Chapter 6

Intercultural Operations

Intercultural operations are mostly depends on the land preparation and crop sowing condition and growth stages (Table 2). If the land preparation is done properly and sowing is done after optimum moisture condition in the soil then only 2 to 3 weeding and 1-2 hoeing are necessary during early stage of crop growth (Plate No. 2).

Table 2. Growth stages of sweet sorghum

Critical stage	Days after sowing
Flower primordial initiation stage	20 to 25
Boot stage	70 to 75
Flowering stage	80 to 90
Grain formation	95 to 100
Physiological fully mature stage	120 to 125

Chapter 7

Irrigation Management

During the land preparation in dryland as well as in irrigated area the preparation of boxes for plantation is done then the water requirement will be less due to the soil moisture conservation. For higher yield and good quality, the protective irrigation should be given if available during following critical growth stages of crop. The critical stages of crop are 1. Flower primordial initiation stage at 25-30 days after sowing, 2. boot stage at 50-55 days after sowing, 3. flowering stage at 76-80 days after sowing, and 4. grain filling stage at 95 to 100 days after sowing. During *kharif* season irrigation should be adjusted according to rainfall distribution and one or two irrigation may be given during dry spell. In summer season, the crop should be irrigated 3-4 times at the interval of 10-15 days according to the crop situation.

1. Plant Protection Measures

For getting good quality fodder and grain, the crop should be protected from pests and diseases. Sweet sorghum mostly affected by shootfly, stemborer, leaf blight and downy mildew (Plate No. 3). If the sowing of this crop is done in the first week of June, then there will be less incidence of shootfly. For the control of shootfly spray, the sweet sorghum crop on seven or eight days after emergence with endosulphan 35 EC at the rate of 350 ml in 250 liters of water for one hectare. The second spray of the same chemical should be given at 10 days after first spray at the 700 ml in 500 liters of water. For the control of Stemborer, spray 30 EC endosulphan at the rate of 750 ml in 500 liters of water on 30 and 40 days after emergence. For controlling leaf blight disease, spray copper oxichloride at the rate of 1250 gm in 500 liters of water when the disease noticed. The subsequent 1-3 spraying should be given at 15 days interval. The occurrence of downy mildew mostly from the infected plants straws and soils. For controlling downy mildew, uproot disease-infected plants before shredding stage and burn them immediately. As well as spray the sorghum crops with 0.2% Dithane M-45 four times starting from 7th days after planting at an interval of a week.

2. Seed Production of Sweet Sorghum

Rabi season is very good for the seed production of sweet sorghum. Sowing should be done in the second fortnight of September. To get maximum good quality seed yield, the seed rate should be 10 kg/ha and spacing should be 45 x 15cm. Other all operations should be followed as per the normal cultivation of sorghum. If you take such type of precaution then the good quality 15 to 20 quintals /ha seed will be obtained.

Chapter **8**

Maturity and Harvesting of Sweet Sorghum

Sweet sorghum start their flowering from 75 to 80 days from planting and fully mature stage generally occur during 125 to 130 days. The maturing of sweet sorghum can be judged by testing the ^{0}Brix content (TSS) of the juice by using hand refractometer / ^{0}brixmeter (Table 3). When the ^{0}Brix of the juice gives 18 to 21 degree reading then we can say that the juice of sweet sorghum is ready to use for the preparation of syrup, gur or alcohol (Plate No. 4). The harvesting of sweet sorghum should be done with the help of sickle and separate earheads, leaves and canes for different purposes. Dry the earheads for separating grains, and the green leaves can be use for the fodder purpose. The dressed canes can be utilized for the crushing to separate juice and bagasse. The clarified juice use for the preparation of syrup, dextrin's, gur or alcohol and the bagasse can be used for cattle as fodder or as a fuel.

Table 3. Effect of different locations on yield and quality of sweet sorghum (at MPKV, Rahuri and MAU, Parbhani)

Constituent	MPKV, Rahuri	MAU, Parbhani
Green stalk yield (t/ha)	40-50	25-30
Juice extractability (%)	35-40	30-35
^{0}Brix (TSS %)	18-22	16-18
Fermentable sugar in the juice (%)	8-10	13-17
Ethanol (L)/100 L fresh juice	6-8	6.5-7.0
Ethanol yield (L/ha)	800-1200	830-1070

The genotypes RSSV 9 (45.3%), RSSV 8 (17.3%), RSSV 13 (39.1%), RSSV 15 (18.1%) and RSSV 16 (12.5%) produced higher percentage of green stalk yield over the check SSV 84. Almost all the genotypes recorded higher grain yield over SSV 84 (1430 kg/ha). RSSV 9 (14822 L/ha) followed by RSSV 13, and RSSV 8 produced the highest juice yield. The higher TSS recorded by the genotypes, RSSV 14, RSSV 15, RSSV 16, RSSV 7, and RSSV 8 was significantly superior to the SSV 84. Minimum reducing sugars i.e. 0.6, 0.65, 0.75 and 0.79 percent were produced in RSSV 14, 15, 16, 7, and 8 respectively, whereas higher non-reducing sugars were observed in RSSV 16 (13.76%), RSSV (13.28%) and RSSV 15 (12.72%). However, maximum total sugar index was noticed in RSSV 9 (15.95) followed by RSSV 16 (12.93) and RSSV 15 (12.83) as compared to SSV 84 (12.21) (Table 4).

Different sweet sorghum genotypes were tested for yield and quality parameters at different locations in India. The average green stalk yield was recorded at higher level at NRCS, Hyderabad followed by Rahuri and Akola. Similar trend was observed in juice yield. The total sugar content was higher at Surat centers sweet sorghum samples followed by Rahuri, Akola and NRCS, Hyderabad (Table 5).

The sweet sorghum variety SSV 84 was observed to be overall superior to other qualifying genotypes in respect of TSS, green stalk yield and quality parameters of juice, syrup and jaggery. The SSV 84 can be utilized either for syrup or jaggery or alcohol production. The juice, syrup and jaggery quality parameters of SSV 84 are presented in Table 6.

Effect of planting date and variety:

Effect of planting date and variety on dry matter yield, percent stalk, stalk moisture, ^{0}brix, fermentable carbohydrate yield and ethanol yield from sweet sorghum is presented in Table 7. This experiment was conducted at Waseca, MN, USA.

Experiment on sweet sorghum was conducted for two years at Waseca, MN, USA during 1987-1988 to study the varietal differences in their total dry matter yields, percent stalk, stalk moisture, ^{0}brix, fermentable carbohydrate yield, ethanol yield

Table 4. Yield and juice quality of sweet sorghum cultivars grown at MPKV, Rahuri in *kharif* 1996.

Genotype	Green stalk yield (t/ha)	Grain yield (q/ha)	Extraction (%)	Juice yield (L/ha)	TSS (°Brix)	Reducing sugars (%)	Non-reducing sugars (%)	Total sugars (%)	Total Sugar Index	Plant height (cm)
RSSV 7	36.2	18.10	29.3	10599	20.2	0.79	8.93	9.72	10.30	315
RSSV 8	41.4	16.50	30.7	12697	19.9	0.97	7.58	8.55	10.86	323
RSSV 9	51.3	18.70	28.9	14822	18.9	1.06	9.70	10.76	15.95	332
RSSV 10	33.3	22.10	28.1	9372	17.1	1.01	9.41	10.51	9.85	333
RSSV 11	35.6	17.00	27.3	9733	14.8	1.05	8.74	10.04	9.77	275
RSSV 12	35.3	18.08	26.5	9362	16.8	1.10	8.16	9.26	8.67	275
RSSV 13	49.1	18.60	25.1	12373	16.0	1.12	7.67	8.79	10.88	304
RSSV 14	26.8	18.70	23.5	6204	21.8	0.62	13.38	14.0	8.79	294
RSSV 15	41.7	17.00	23.0	9599	21.8	0.65	12.72	13.37	12.83	306
RSSV 16	39.7	18.70	22.4	8913	21.0	0.75	13.76	14.51	12.93	313
SSV 84	35.3	14.30	28.4	10035	19.0	0.94	11.23	12.17	12.21	293
S.E. ±	2.10	1.80	2.65	2171	2.27	0.17	2.17	2.02	2.05	19
CD at 5 %	5.80	5.42	7.98	6516	6.82	0.52	6.53	6.07	6.16	58

Table 5. Yield and quality of sweet sorghum genotypes grown at different locations in India

Genotype	Location	No. of year tested	Green stalk yield (t/ha)	Grain yield (q/ha)	Juice extraction (%)	Juice yield (L/ha)	TSS (⁰Brix)	Reducing sugar (%)	Non-reducing sugar (%)	Total sugar (%)	Total sugar index	Plant height (cm)
GSSV 148	Surat	3	37.1	18.25	23.3	5403	16.1	3.12	11.78	14.90	8.05	258
GSSV 153	Surat	3	32.7	28.68	18.5	1546	17.1	2.49	11.77	14.26	2.20	231
GSSV 236	Surat	2	42.3	38.88	28.3	10874	19.6	1.11	9.49	10.60	11.53	263
SR 350-3	Surat	3	33.1	28.27	19.1	4554	16.6	1.06	8.40	10.06	4.60	262
AKSS 15	Akola	3	36.8	24.12	29.4	10027	16.9	1.11	9.87	10.98	11.00	297
AKSS 16	Akola	4	42.1	17.37	27.9	14018	18.2	0.87	9.92	10.79	15.13	295
AKSS 18	Akola	3	38.5	20.81	27.5	6312	16.6	2.49	11.47	13.96	8.81	270
RM 47	NRCS, Hyderabad	2	41.0	17.23	28.6	12040	15.9	1.12	6.56	7.68	9.25	277
SSH 1	NRCS, Hyderabad	4	41.4	31.74	28.6	9525	15.5	1.09	7.86	8.25	8.57	287
RS 608	NRCS, Hyderabad	2	46.0	17.78	29.1	12503	16.4	0.84	6.38	7.22	9.10	293
NSS 102	NRCS, Hyderabad	4	37.6	29.67	26.7	10356	16.5	0.85	8.90	9.75	10.09	292
SSV 84	MPKV, Rahuri	5	39.9	16.96	30.2	11543	17.6	1.25	10.27	11.52	13.30	277
S.E. ±	-	-	3.71	6.92	3.79	3597	1.06	0.74	1.77	2.41	3.33	18.73
CD at 5 %	-	-	11.13	20.76	11.39	10792	3.21	2.24	5.31	7.25	10.00	56.20

Table 6. Quality of juice, syrup and jaggery prepared from sweet sorghum *var.* SSV 84.

Quality parameter	Juice	Syrup	Jaggery
Moisture (%)	-	32.00	10.93
T.S.S. (^{0}Brix)	20.23	68.00	-
pH (6.5 % solution)	4.86	5.1	4.9
Acidity (%)	0.45	2.05	2.28
Colour of 6.5 % solution (O.D. value)	-	0.500	0.954
Reducing sugars (%)	1.90	6.48	10.62
Non-reducing sugars (%)	12.43	30.38	37.73
Total sugars (%)	13.33	36.86	48.36
Starch (%)	0.72	-	-
C.C.S. (%)	6.09	-	-
C.C.S. (q/ha)	24.68	-	-

Table 7. Effect of planting date and variety on dry matter yield, percent stalk, stalk moisture, ^{0}brix, fermentable carbohydrate yield and ethanol yield of sorghum harvested at Waseca, MN. USA (Harvesting mid October).

Date of planting	Total dry matter (t/ha)	Per cent stalk[1] (%)	Stalk moisture (%)	^{0}Brix	Fermentable carbohydrate yield[2] (t/ha)	Calculated ethanol yield[3] (gal/a)
April 26	9.3	66	68	11.3	1.71	232
May 5	8.6	66	69	10.8	1.60	217
May 15	8.4	65	69	10.4	1.46	199
May 25	8.2	66	70	10.4	1.52	206
LSD (0.05)	1.0	2	1	0.6	0.18	24
Variety mean						
8361 A	9.2	70	70	9.1	1.49	101
405	8.7	69	67	9.2	1.26	86
301	8.2	53	68	11.4	1.22	83
Keller	8.4	71	70	13.2	2.32	158
LSD (0.05)	0.8	2	1	0.6	0.18	12

and stalk lodging (Table 8). The primary purpose of this research was to evaluate sweet and semisweet sorghum as a feedstock for ethanol production. Keller, Dale and M81E competed favorably with corn in theoretical ethanol production based on amount of sugar produced. However, these varieties lodged severely both years, which would have made mechanical harvesting difficult.

1. Dry matter basis.
2. ⁰Brix multiplied times stalk sap yield.
3. Assuming 2.5 gal of ethanol/bu of corn grain and 14.7 lb fermentable carbohydrate/gal of ethanol.

Table 8. Varietal differences in total dry matter yields, per cent stalk, stalk moisture, ⁰brix, fermentable carbohydrate yield, ethanol yield and stalk lodging of sorghum and corn grown at Waseca, MN, USA (average of two years)

Genotype	Total dry matter (t/ha)	Per cent stalk[1] (%)	Stalk moisture (%)	⁰Brix[2]	Fermentable carbohydrate yield[3] (t/ha)	Calculated ethanol yield[4] (gal/a)	Stalk lodging[5] (%)
Corn							
Dekalb 524	8.2	-	46	-	-	332	15
Sorghum							
Northrup king 301	9.4	52	68	13.2	1.81	247	47
Rox orange	10.6	46	75	10.3	1.84	250	33
Northrup king 405	11.0	60	70	7.3	1.25	170	70
Keller	10.1	70	72	13.4	2.96	403	97
Dale	10.1	70	74	12.3	3.00	408	98
Northrup king 8361 A	11.7	67	70	8.9	1.85	252	72
M 81E	10.0	66	73	12.7	2.83	385	96
Northrup king 8361	13.1	68	73	7.4	1.96	267	93
LSD (0.05)	1.5	3	3	1.7	0.49	66	18

1. Dry matter basis.
2. ⁰Brix is approximately equivalent to percent sugar.
3. ⁰Brix multiplied times stalk sap yield.

4. Assuming 2.5 gal of ethanol/bu of corn grain and 14.7 lb fermentable carbohydrate/gal of ethanol.
5. Stalk lodging percent of plants lodged at least 450 or more at harvest.

The higher green cane yield was obtained in RSSV 47 (44.9 t/ha) followed by NSS 104 (42.2 t/ha) in 1st planting (June). The green cane yield was highest in June planting followed by December, April and February. The genotypes NSS 104, RSSV 9 and RSSV 47 had high green cane yield and suitable for June, December, February and April planting (Table 9). The highest grain yield was obtained in 1st planting i.e., in June. Overall, the hybrid NSS 104 recorded highest grain yield (10.9 q/ha). The grain yield ranges from 2.1 q/ha from Keller (April planting) to 10.9 q/ha from NSS 104 (June planting) (Table 9).

In June panting, the genotypes RSSV 47 (28.9%) and RSSV 9 (28.1%) showed highest juice extraction percentage. Wray (21.5%) and NSS 104 (18.5%) recorded highest juice extraction percentage in August planting. In October, planting RSSV 47 (36.1%) recorded highest juice extraction percentage followed by NSS 04 (32.0%). The genotypes AKSS 16 (27.1%) and Keller (23.5%) showed highest juice extraction in December planting. In February, planting Wray (24.5%) and RSSV 9 (24.1%) showed maximum juice extraction. The sweet sorghum planting in April the genotypes Wray (25.5%) and RSSV 47 (23.7%) showed highest juice extraction percentage. Overall, the genotypes RSSV 9, RSSV 47, NSS 104, Wray and Keller showed higher juice extraction percentage. In all, the genotypes Wray, RSSV 9, NSS 104 and RSSV 47 showed higher ^{0}brix percentage than the rest of the genotypes and it ranges from 15.2 to 20.3% (Table 10).

The non-reducing sugars content in sweet sorghum genotypes planted in different months were ranged from 10.3 to 17.5%. The maximum non-reducing sugars content was observed in Wray (17.5%) in December planting followed by RSSV 9 in April and February planting (17.2%). Overall, Wray, RSSV 9 and Keller showed higher non-reducing sugars content in almost all the sowing dates (Table 11). The highest total sugars content was observed in Wray (18.2%) in December, planting followed by RSSV 9 in February and April planting (18.2% and 18.1%, respectively). In all, the genotypes Wray, Keller, NSS 104, RSSV 9 and RSSV 47 showed higher total sugars content than the rest of the genotypes (Table 11).

The biomass production of sweet sorghum was highest in June planting season followed by December planting. The highest biomass production observed in RSSV 47 (66.2 t/ha) during June planting followed by NSS 104 (63.5 t/ha), and RSSV 9 (61.2 t/ha). The juice yield from sweet sorghum is mostly depending upon the actual green cane yield and juiciness of the cane. The juice yield from sweet sorghum ranged from 1488 to 12976 L/ha (Table 12). Among the six different planting dates June planting of sweet sorghum gave highest recovery of juice. The highest juice recovery was recorded in RSSV 47 (12976 l/ha) followed by RSSV 9 (11212 L/ha) and SSV 84 (10073 L/ha).

Table 9. Effect of sowing dates on yield and juice quality of sweet sorghum.

Genotype	Cane yield (t/ha)						Grain yield (q/ha)					
	June	Aug.	Oct.	Dec.	Feb.	April	June	Aug.	Oct.	Dec.	Feb.	April
RSSV 9	39.9	24.9	24.4	34.5	27.1	32.4	10.3	4.9	4.9	4.7	4.1	4.5
RSSV 47	44.9	19.0	24.5	33.7	28.6	31.0	10.7	6.4	5.7	5.6	3.5	4.3
NSS 04	26.8	20.3	20.0	29.1	20.5	21.9	10.0	6.9	5.0	4.9	4.6	3.3
NSSV 104	42.2	25.5	26.3	36.3	19.4	32.7	10.9	6.9	5.1	5.1	5.4	4.9
Madhura	23.6	19.0	13.4	19.9	16.5	20.0	5.7	4.4	3.9	4.1	4.3	3.5
BJ 248	27.0	14.5	20.9	22.8	15.4	15.9	7.2	3.8	4.7	2.4	5.1	4.5
Wray	28.0	19.2	21.0	23.0	18.0	25.9	6.4	4.2	4.3	2.5	3.6	3.5
Keller	28.5	25.2	27.9	29.2	23.4	35.6	5.9	4.5	4.3	2.9	5.1	2.1
AKSS 16	18.7	14.8	17.6	25.3	20.5	26.3	5.7	3.8	3.5	4.0	4.1	4.4
HES 04	25.1	12.1	12.8	29.2	18.6	19.6	7.7	4.7	3.6	4.7	4.0	4.4
SSV 84	38.3	10.5	19.8	28.3	23.6	28.4	8.3	4.6	3.9	5.4	4.0	4.3
S.E. ±	2.58	2.14	2.28	3.65	1.06	1.72	0.50	0.20	0.20	3.77	0.30	0.30
C.D. at 5%	7.62	6.33	6.74	10.77	3.15	5.08	1.60	0.70	0.50	11.13	1.00	0.90

Table 10. Effect of sowing dates on yield and juice quality of sweet sorghum.

Genotype	Juice percentage (extraction percentage)						^{0}Brix of the juice					
	June	Aug.	Oct.	Dec.	Feb.	April	June	Aug.	Oct.	Dec.	Feb.	April
RSSV 9	28.1	14.3	17.9	20.8	24.1	19.2	18.2	17.1	18.8	18.6	19.7	19.5
RSSV 47	28.9	18.1	36.1	22.6	22.4	23.7	19.3	18.1	18.3	17.9	18.7	18.8
NSS 04	25.5	15.3	32.0	19.8	20.3	18.8	17.8	16.9	16.5	16.3	15.3	15.2
NSSV 104	22.4	18.5	21.3	20.2	21.8	22.5	19.2	18.6	16.8	17.6	17.6	18.3
Madhura	23.2	18.1	34.9	21.8	19.1	20.5	17.6	16.1	17.8	18.1	18.6	17.8
BJ 248	20.1	15.7	24.5	16.6	13.0	14.6	17.9	16.1	16.4	16.4	16.9	16.5
Wray	26.7	21.5	16.5	21.0	24.5	25.5	20.3	18.5	18.4	19.5	21.5	17.5
Keller	15.4	15.4	18.3	23.5	23.1	22.5	19.7	17.5	18.2	19.0	19.1	19.7
AKSS 16	18.7	18.7	23.1	27.1	19.9	22.8	18.3	17.1	15.7	16.5	17.0	16.8
HES 04	20.0	12.3	13.7	22.6	11.9	16.7	16.4	16.6	16.1	15.9	16.1	16.3
SSV 84	26.3	16.0	20.3	22.5	23.8	23.8	18.1	16.8	17.4	16.9	16.7	17.3
S.E. ±	1.92	1.41	2.18	1.43	1.77	1.63	0.27	0.35	0.30	0.27	0.32	0.41
C.D. at 5%	5.68	4.17	6.40	4.22	5.22	4.82	0.80	1.03	0.79	0.79	0.94	1.19

Table 11. Effect of sowing dates on yield and juice quality of sweet sorghum.

Genotype	Non-reducing sugars (%)						Total sugars (%)					
	June	Aug.	Oct.	Dec.	Feb.	April	June	Aug.	Oct.	Dec.	Feb.	April
RSSV 9	14.9	13.8	15.0	16.9	17.2	17.2	16.6	14.9	16.1	17.5	18.2	18.1
RSSV 47	16.2	13.9	15.6	16.4	16.2	16.0	17.5	15.1	16.8	17.5	17.3	17.1
NSS 04	13.8	11.5	13.6	13.8	12.1	12.4	15.6	13.0	15.0	15.1	13.6	13.7
NSSV 104	16.6	14.2	13.7	15.0	15.8	14.1	19.8	15.4	15.3	16.3	17.1	15.2
Madhura	14.3	12.2	15.4	16.1	14.3	15.7	15.7	13.2	16.5	16.9	15.4	16.6
BJ 248	14.2	13.1	11.6	13.3	14.3	13.1	16.0	14.9	13.2	14.7	15.9	14.8
Wray	14.5	15.5	16.3	17.5	16.9	15.7	16.5	16.5	17.2	18.2	17.9	16.6
Keller	17.0	14.2	16.1	16.4	16.2	16.8	7.7	15.2	17.1	17.2	17.1	17.6
AKSS 16	14.9	12.7	11.7	13.8	14.1	12.9	6.6	14.2	13.3	15.3	15.8	14.4
HES 04	15.3	10.3	12.5	13.3	13.8	13.2	16.5	12.2	14.2	14.9	15.2	14.9
SSV 84	14.7	14.7	13.8	14.7	14.1	14.7	15.9	14.2	15.3	16.1	15.3	16.0
S.E. ±	0.99	1.44	1.61	1.45	1.49	1.59	0.16	0.09	0.09	0.07	0.21	0.14
C.D. at 5%	2.98	4.33	4.84	4.37	4.47	4.79	0.48	0.26	0.26	0.26	0.62	0.43

Table 12. Effect of sowing dates on yield and juice quality of sweet sorghum.

Genotype	Biomass (t/ha)						Juice yield (L/ha)					
	June	Aug.	Oct.	Dec.	Feb.	April	June	Aug.	Oct.	Dec.	Feb.	April
RSSV 9	61.20	46.22	45.72	55.81	48.42	53.72	11212	3561	4368	7176	6531	6221
RSSV 47	66.20	40.31	45.82	55.01	49.92	52.31	12976	3439	8845	7616	6406	7347
NSS 04	48.11	41.62	41.31	50.42	41.82	43.22	6834	3106	6400	5762	4162	4117
NSSV 104	63.52	46.82	47.62	57.62	40.72	54.01	9453	4718	5602	7333	4229	7358
Madhura	44.91	40.41	34.72	41.22	37.83	41.32	5475	3439	4677	4338	3152	4100
BJ 248	48.31	35.82	42.22	44.11	36.71	37.22	5427	2277	5121	3785	2002	2321
Wray	49.31	40.52	42.32	44.31	39.31	47.22	7476	4128	3465	4830	4410	6605
Keller	49.82	46.52	49.22	50.51	44.70	51.60	4389	3881	5106	6862	5405	8010
AKSS 16	40.01	36.12	38.93	46.61	41.81	47.61	3497	2768	4066	6856	4080	5996
HES 04	46.63	33.52	34.16	50.51	39.91	40.92	5080	1488	1754	6599	2213	3273
SSV 84	59.63	31.83	41.12	49.61	44.92	49.72	10073	1680	4019	6368	5617	6759
S.E. ±	8.19	4.97	4.64	4.97	4.01	5.48	2932	958	1709	1229	1456	1794
C.D. at 5%	24.57	14.92	13.93	14.91	12.03	16.44	8796	2876	5128	3688	4369	5385

Chapter 9

Effect of Delay in Harvesting and Storage Period on Cane Yield and Juice Quality

Sweet sorghum canes from RSSV-9 were harvested at a time and twenty-one bundles were prepared containing ten canes in each bundle and kept in open environment. Every day three bundles of ten canes were used for observation and extraction of juice as well as studying juice quality parameters. Within seven days of storage period of sweet sorghum stalks in sunlight loss of 22.23% in weight while the juice extraction percentage also reduced upto 4.89% (Table 13). Due to reduction in weight of the cane the ^{0}Brix was increased from 21.2 to 28.0 ^{0}Brix. There was slight change in PH of the juice. As the period of storage increased, the percentage of reducing sugars also increased as well as the total sugars also increased but the rate of increase in the non-reducing sugars was decreased. These results indicate that the harvested sweet sorghum canes must be processed as early as possible to avoid the losses.

Sweet sorghum canes from RSSV-9 were harvested at three different time intervals. 1. At zero days' interval (at physiological maturity stage), 2. fifteen days after physiological maturity and 3. thirty days after physiological maturity. Delay in harvesting of sweet sorghum after physiological maturity decline the juice extraction percentage. Within one month it reduces juice extraction upto 7% (Table 14), but due to loss in water content from the sweet sorghum cane increases ^{0}Brix as well as reducing and total sugars. These results also indicated that if sweet sorghum harvesting delayed beyond the physiological maturity date it decline the juice extraction as well as the quality of the juice.

Table 13. Effect of storage period on harvested sweet sorghum cane and juice.

Days after harvest	Loss in wt. (%)	Juice extraction (%)	Loss in juice extraction (%)	⁰Brix	pH	Reducing sugars (%)	Non-reducing sugars (%)	Total sugars (%)
0	-	16.52	-	21.2	6.0	0.88	9.78	10.66
1	4.17	14.81	1.71	21.8	6.0	0.92	11.56	12.48
2	9.86	13.33	3.19	24.0	6.2	0.99	12.70	13.69
3	15.87	12.82	3.70	26.0	6.3	1.08	12.96	14.04
4	20.13	12.30	4.22	26.4	6.3	1.83	13.94	15.77
5	21.49	12.20	4.32	27.0	6.3	2.81	13.14	15.95
7	22.23	11.63	4.89	28.0	6.3	3.19	13.10	16.29
S.E. ±	6.62	1.60	1.02	2.44	0.13	0.90	1.27	1.91
CD at 5 %	19.86	4.79	3.07	7.31	0.40	2.69	3.82	5.74

Table 14. Effect of delay in harvesting of sweet sorghum on juice quality.

Harvesting period (days) after physiological maturity	Juice extraction (%)	Loss in juice extraction (%)	⁰Brix	pH	Reducing sugars (%)	Non-reducing sugars (%)	Total sugars
0	16.52	-	21.2	6.0	0.88	9.78	10.66
15	10.47	6.05	22.0	6.5	1.85	11.03	12.88
30	9.53	6.99	23.0	6.3	2.56	11.03	13.59

Study on the influence of stage of harvesting on changes in quality, stalk yield and biomass in sweet sorghum cultivars was under taken with 4 different genotypes. The biomass production was highest when harvested at 15 days after 50% flowering (soft dough stage) in SSV 84, CSV 19SS and PAC 52093 but it was highest at harvesting 30 days after 50% flowering (hard dough stage) in CSH 22SS (Table 15). The cane yield ranged from 22.76 to 44.85 q/ha. It was highest in CSH 22SS harvested at 30 days after 50% flowering (hard dough stage) (44.85 t/ha) followed by CSV 19SS (43.75 t/ha) harvested at 15 days after 50% flowering (soft dough stage), SSV 84 (31.56 t/ha) harvested at 15 days after 50% flowering and PAC 52093 (31.40 t/ha) harvested at 50% flowering. As the harvesting period from 50 % flowering increases the juice extraction percentage decreases. The highest juice extraction was at 50% flowering stage in CSV 19SS (34.59 %) followed by CSH 22SS (31.00%), PAC 52093 (28.80%) and SSV 84 (24.67%). The juice yield was highest at harvesting 15 days after 50% flowering (soft dough stage) in CSV 19SS (12314 L/ha), PAC 52093 (8937 L/ha) and SSV 84 (7310 L/ha), but in CSH 22SS (12432 L/ha) was highest when sweet sorghum harvested at 50% flowering stage. As the harvesting period increases, the ^{0}Brix also increases from 9.3 to 20.5. The highest ^{0}Brix reading was in SSV 84 (20.5) and PAC 52093 (20.5) followed by CSH 22SS (20.3) and CSV 19SS (20.0) at physiological maturity stage. There was no much variation in the pH of the juice. There was no specific trend in the reducing sugar content. The non-reducing and total sugars in all varieties increased as the harvesting period increases. The highest accumulation of sugars was observed at physiological maturity stage. Highest total sugar content was observed in SSV 84 (11.32%), followed by PAC 52093 (10.32%), CSH 22SS (9.80%) and CSV 19SS (9.50%). The CCS production was highest when sweet sorghum harvested at physiological maturity stage. It was highest in SSV 84 (12.32 q/ha) followed by CSH 22SS (11.83 q/ha), PAC 52093 (10.40 q/ha) and CSV 19SS (10.11 q/ha). The total sugar index ranged from 2.32 to 7.62. The highest TSI was found in CSH 22SS (7.62) followed by PAC 52093 (7.48), CSV 19SS (5.73) and SSV 84 (4.56).

Considering the above studied yield and juice quality parameters of sweet sorghum harvested at 30 days after 50% flowering (hard dough stage) is appropriate to get desired ^{0}Brix, total sugars and TSI. The entry CSH 22SS was found promising followed by CSV 19SS, PAC 52093 and SSV 84. Mature sweet sorghum canes shown in Plate No. 5.

Table 15. Effect of different harvesting stages on yield and quality of sweet sorghum

Genotype / harvesting stage	Biomass yield (t/ha)	Cane yield (t/ha)	Juice yield (L/ha)	°Brix (%)	pH	Reducing sugars (%)	Non-reducing sugars (%)	Total sugars (%)	C.C.S. (%)	C.C.S. (q/ha)	TSI	Juice extraction (%)
A. SSV 84												
1. 50 % flowering	36.46	22.76	5623	9.5	6.53	0.99	3.15	4.14	0.45	1.03	2.32	24.67
2. Soft dough stage	53.50	31.56	7310	14.0	6.97	1.33	6.23	7.56	2.30	6.88	5.42	23.11
3. Hard dough stage	47.34	26.44	5396	17.8	6.53	0.91	6.93	7.84	1.89	4.91	4.22	20.48
4. Fully mature stage	46.36	26.75	4029	20.5	6.57	1.06	10.26	11.32	4.56	12.32	4.56	15.09
B. CSH 22SS												
1. 50 % flowering	59.48	39.83	12432	9.3	6.80	1.32	2.57	3.89	0.24	0.98	4.74	31.00
2. Soft dough stage	69.09	41.83	10762	14.2	7.13	1.67	4.71	6.38	0.68	2.88	6.91	25.61
3. Hard dough stage	73.75	44.85	10501	18.0	6.76	0.93	6.09	7.02	0.99	4.43	7.37	23.43
4. Fully mature stage	64.49	39.36	7725	20.3	6.83	1.16	8.72	9.88	3.02	11.83	7.62	19.68
C. CSV 19SS												
1. 50 % flowering	58.06	34.75	12035	10.0	6.77	0.78	2.66	3.44	0.40	1.43	4.10	34.59
2. Soft dough stage	74.88	43.75	12314	14.7	7.23	0.97	5.21	6.18	1.06	4.66	7.62	28.06
3. Hard dough stage	70.41	41.96	8929	17.8	6.80	0.88	5.92	6.80	0.86	3.51	6.06	21.31
4. Fully mature stage	64.41	37.72	6063	20.0	6.83	1.12	8.38	9.50	2.77	10.11	5.73	16.53
D. PAC 52093												
1. 50 % flowering	51.10	31.40	8794	10.8	6.83	0.76	4.67	5.43	1.64	5.31	4.82	27.67
2. Soft dough stage	51.91	30.95	8937	14.8	7.27	0.89	6.10	6.99	1.93	6.01	6.26	28.80
3. Hard dough stage	43.58	26.20	7115	18.5	6.77	0.83	6.54	7.37	1.31	3.39	5.25	27.18
4. Fully mature stage	48.16	29.77	7198	20.5	6.77	1.16	9.16	10.32	3.42	10.40	7.48	25.25

Chapter **10**

Utilization of Sweet Sorghum as a Fodder

Fodder quality of sweet sorghum:

The fodder quality of sweet sorghum is mostly depends upon the chemical constituents mostly present in the sweet sorghum stover. Various fodder quality parameters of sweet sorghum are presented in Table 16.

Table 16. Fodder quality of sweet sorghum.

Quality parameter	Per cent content (range)
Dry matter	80-82
Total soluble solids	7.5-8.5
IVDMD	54-65
Proteins	5.5-6.5
Crude fiber	27-32
Total ash	8.5-9.5
Nitrogen free extract	50-55
HCN (ppm)	70-150
Voluntary intake	75-85

India rank first in the world for the milk production. This achievement is possible due to the raring of cross bread cows, buffalos, and sheeps on a large numbers. In India, most of the land cultivation practices are carried out with the help of bullocks. For that purpose farmers are raring good quality Khilar breeds. For maintaining these all cattle heard in good condition to get more work done and more milk production, we have to provide sufficient quantity and quality fodder material to them. The higher milk product-producing animal yields more milk when fed with nutritive and succulent feed in required quantities. This indicates that there is a direct relationship between the milk production and good quality feed/fodder supply to the animals. Due to increased pressure on arable land for grain and commercial crop and heavy pressure of human population, it is not possible to extend area under fodder crops. One of the options is to raise high yielding forage genotypes. Sorghum for forage is grown under both rainfed and irrigated conditions. Almost 60 to 70 per cent of total forage demand in *kharif* season is met from sorghum. During *kharif* season growing of hybrid sorghum produce some quantity of fodder but due to the poor quality and acceptability milch animals are not eating sufficient quantity, due to this reason the milk production goes down by feeding such type of fodder. Therefore, sweet sorghum cultivation is specially recommended as a fodder crop for the milch animals. If you see the nutritive value of sweet sorghum (Table 17), it indicates that animals are eating this fodder very efficiently. New genotypes of sweet sorghum grown in *kharif* or Summer season gives 45 to 50 tones/ha green fodder yield within 90 days. If same crop is kept for ratton, it gives 30-35 tones/ha green fodder yield. On the basis of all characters of sweet sorghum, it is concluded that the cultivation of sweet sorghum in *kharif* or summer season for milch animals or other animals too, it gives very good quality and quantity of fodder.

Table 17. Fodder quality of sweet sorghum

Quality parameter	Percentage
Dry matter	80.20
Protein	6.58
IVDMD	54.8
TSS	8.34
Crude fiber	27.97
Total ash	9.45
Nitrogen free extract	54.47
HCN (ppm)	73.00

Sorghum Fodder Processing (Sorghum fodder bricks)

Animals need dry fodder for their well development as well as good health and milch animals for higher milk production. Due to feeding dry fodder to the animals they chew it during resting period. While chewing their feed, saliva as well as other digestive enzymes mix in the feed and help for the digestion. Due to this development ultimately the milk production increase. In Maharashtra Solapur, Satara, Sangli, Kolhapur, Pune and Thane are the districts uses dry sorghum fodder as a main dry fodder to the animals. During famine, very heavy rainfall and during other worst condition sorghum dry fodder survives the animals. For making available dry sorghum fodder all over the year it needs to be converted into a bricks or other form to store in a suitable storage condition.

For milch animals fodder and feeds have very high importance. Now a days most of the milk producer gave tablet form feeds to the animals. These tablet form feeds are manufacture from cotton seed cake, groundnutcake, safflower cake etc. Some times sunflower, salt, urea, maize, wheat straw, sorghum, bajara, and molasses are used.

Recently Mangalweda Kadaba-Dryfodder Research and Development Corporation have started to prepare sorghum stover bricks manufacturing by using hydrolic press machine. In this process, they are cutting sorghum fodder and then mixing with molasses and other plants straw such as maize straw, gram straw, pigeon pea straw, groundnut straw, wheat straw, minerals and vitamins also added in the sorghum stover bricks (Plate No. 6). Therefore, their nutritional value is increasing.

Due to such type of fodder and feeds are prepared from plant origin and that are easily available to the farmers. By feeding such type of fodder bricks they can reduce expenditure on other feeds.

For storage of sorghum stover needs large space as well as it remain in open condition. Due to this situation during rainy season, bad climate, sorghum stover as well as other stovers get damaged. Damaged sorghum stover will not eat by animals.

If we prepare such type of sorghum fodder bricks, they can be store under shed in minimum space and make it available throughout the year with improved nutritional quality. This will give the additional benefits to the farmers to use straws from other plant sources with sorghum stover for preparation of fodder bricks. Therefore, farmers have to come forward to take advantage of such facilities from the above firm.

Chapter 11

Cost of Cultivation of Sweet Sorghum

The data obtained from two industries viz., Sagar Sugar and allied Chemicals Ltd. (Chittoor) and Praj Industries Ltd. (Pune), the paid-out cost of sugarcane is 2.3 to 3.6 times that of incurred in sweet sorghum cultivation. The gross income received is to the tune of Rs. 30000 - 34000 per hectare from sweet sorghum cultivation, which is bringing farmers only 47 - 53 per cent of gross income realized from sugarcane cultivation. However, the higher gross income from sugarcane should be suitably discounted to compare it with that of sweet sorghum on crop duration basis. Sugarcane harvested in 12 months whereas sweet sorghum takes only 4 months to yield. The out put-in put ratio at paid out cost is for sweet sorghum high compared to sugarcane in both trials. It can be interpreted that sweet sorghum cultivation can yield proportionately higher returns per rupee invested in a shorter crop period (Table 18, Plate 6, 7). Though it is agreed that shorter duration of sweet sorghum

Table 18. Comparison on cost of cultivation of sweet sorghum and sugarcane at two different locations in India.

Sr. No	Operation	Maharashtra		Andhra Pradesh	
		Sweet sorghum	Sugarcane	Sweet sorghum	Sugarcane
		(Cost, Rs/ha)		(Cost, Rs/ha)	
1.	Land preparation	1875	3750	1875	2500
2.	Seed cost	875	3750	320	8750
3.	Sowing charges	625	-	625	-
4.	Planting	-	3750	500	2000
5.	Weed control	1250	2500	500	2000

6.	Inter cultivation	-	-	1250	1500
7.	Earthing-up	-	500	-	500
8.	Manures and fertilizers	2500	7500	2500	6250
9.	Irrigation-labour and electricity	500	12500	500	2500
10.	Labour charges	1250	7500	1000	7500
11.	Harvesting and transport charges	5000	7000	8750	15750
12.	Paid-out cost	13375	48750	17820	49250
		Returns (Rs/ha)			
1.	Cane yield (t/ha)	50	75	50	75
2.	Cane price (Rs/t)	400	850	400	850
3.	Grain yield (q/ha)	20	-	20	-
4.	Cost of grain (Rs/kg)	5	-	7	-
5.	Gross income (Rs/ha)	30000	63750	34000	63750
6.	Paid-out cost	13375	48750	17820	49250
7.	Net income at paid-out cost	16625	15000	16180	22000
8.	Output-input ratio at paid-out cost	1.80	1.32	1.91	1.53

Sources: Sagar Sugars and allied chemicals chittor.
Praj Industries Ltd., Pune.

allows us to take up two or even three crops in a year, the yield level/potential of the *Kharif* sweet sorghum cannot be replicated in the other two crop seasons.

Chapter 12

Crushing of Sweet Sorghum Canes for Juice Extraction

The dressed sweet sorghum canes should be crushed within 6 to 12 hrs to get good quality juice. If there is delay in cane crushing start the stailing effect on the juice quality. Delay in crushing start inversion of the sucrose into glucose and that affect the quality of syrup, gur and other products. After harvesting sweet sorghum canes should be crushed with the help of cane crusher having three stainless steel rollers in a horizontal position (Plate No. 8). While crushing the cane and extracting juice, if the juice come in the content of iron metals it turns to black as well as some phenolics and chlorophyll extract in the juice. Therefore care should be taken to avoid any contamination and change in the juice colour. The storage of juice should be done in the stainless steel container or cement tank to avoid change in pH of the Juice (Plate No. 9). After extracting sufficient quantity of juice, it should be transferred for further process immediately.

Clarification of juice for good quality:

From the results, it was observed that after centrifugation and addition of α- amylase (50 ppm) there was no change in TSS and pH. However, there is slight reduction in acidity in terms of aconitic acid. Acidity was reduced from 353 mg/100g to 319 mg/100g of juice (Table 19). It is interesting to note that there was spectacular reduction (87.24%) in the starch content of sweet sorghum juice (Table 20). However, there was increase in reducing sugars by 17.16% was observed. It resulted in reduction of non-reducing sugars by 5.27% as a result of some inversion might be taken place in non-reducing sugar. Considering above observations, it is possible to remove the starch from sweet sorghum juice to the extent of 87.24% by centrifugation and addition of α- amylase (50 ppm) prepared from germinating barley seeds.

Table 19. Effect of centrifugation and addition of α-amylase on removal of starch and on juice quality

Parameter	Before centrifugation and addition of α-amylase	After centrifugation and addition of α-amylase
Starch (%)	0.768	0.098
T.S.S. (^{0}Brix)	20.00	20.00
PH	4.90	4.90
Acidity (mg/100 g)	353.00	319.00
Reducing sugars (%)	1.28	1.50
Non-reducing sugars (%)	9.29	8.80
Total sugars (%)	10.57	10.30

Table 20. Effect of centrifugation and addition of α-amylase in sweet sorghum juice in percent.

Parameter	Per cent
Reduction in starch	87.24
Increase in reducing sugars	17.18
Reduction in non-reducing sugars	5.27
Reduction in total sugars	2.55

During *kharif* season of 1985 and 1986 twelve sweet sorghum varieties were tested for juice quality. The juice of these varieties showed pH range from 4.9 to 5.3. The cultivar SSV 7073 recorded the highest ^{0}brix (22.24%) with better Pol percentage juice (15.87%). The TSS of all the cultivars ranged from 10.00 to 22.5% from flowering to maturity. Minimum reducing sugars (0.73%) was observed in the cultivar SSV 1333 with better extraction (37.45%). Considering all the quality parameters the cultivar SSV 1333 was found superior amongst all the cultivars tested (Table 21).

Studies on juice quality during storage

A study on changes of the chemical quality of juice was done in terms of its inversion at different periods of time after harvesting, Juice ^{0}Brix, pH, total sugars, reducing sugars and sucrose were evaluated. Cane from NSSH 104 was harvested and upto 48 hours with an interval of 6 hours, samples were drawn and juice quality was determined. After 48 hours, 17% reduction in sucrose content was observed. This is the percent inversion of sucrose after 48 hours (Table 22).

Boiling and Removing Molasses from Juice

After crushing sweet sorghum canes and getting sufficient quantity of juice (750 liters) immediately take that juice in to the metal pan having inner layer with zinc and already kept on the fire box while pouring juice into the boiling pan filter with nylon or muscling cloth. Then start the heating of juice slowly, while boiling of juice adds 150-200g Calcium hydroxide already dissolved in the 5 liters of water.

At the same time or at the time of boiling of juice add 1.5kg pest of okra (stem + branches) plant, which is prepared separately. This will help to separate all molasses from the juice and juice become clear. Due to the addition of calcium hydroxide it coagulate nitrogenous substances and come above/top of the juice with mucilage or molasses. This should be removed from the juice to get clear juice for the preparation of syrup or jaggery. After removing molasses from the juice add 150 to 200 ml phosphoric acid for 1000 liters to neutralize the excess amount of calcium hydroxide as well as to avoid inversion of the sucrose. While adding phosphoric acid, the care should be taken to avoid excess amount because if it become excess then the setting of jaggery will not occur and jaggery remain as semi-solid.

Table 21. Yield and juice quality of sweet sorghum genotypes

Genotype	Green stalk yield (t/ha)	Grain yield (q/ha)	Commercial cane sugars (t/ha)	Juice extraction (%)	pH	TSS (°Brix)	POL % of juice	Purity (%)	POL % of cane	Reducing sugars (%)	Millable cane height (cm)
SSV 74	52.22	19.54	4.77	42.60	5.1	19.96	13.61	68.17	7.79	1.42	276
SSV 84	42.18	15.42	3.24	31.80	5.1	21.56	13.98	64.84	7.68	1.68	259
SSV 108	41.51	41.05	3.78	25.60	5.0	22.04	15.61	70.83	9.19	1.24	260
SSV 111	32.92	32.12	2.36	36.93	5.2	18.34	12.50	68.16	7.18	1.99	302
SSV 119	39.97	22.63	3.41	38.95	4.9	19.09	14.05	73.60	5.82	1.39	295
SSV 685	36.98	18.98	3.07	32.07	4.9	19.06	13.93	73.04	8.40	1.33	275
IS 956	28.47	39.09	2.37	27.31	5.2	20.01	14.14	70.66	8.33	1.28	318
SSV 1333	33.69	19.95	3.51	37.45	5.0	21.46	16.65	77.59	10.41	0.73	260
SSV 2525	57.64	22.63	4.45	29.90	5.0	19.56	13.42	68.61	7.72	1.94	344
SSV 3555	30.53	20.57	3.13	28.20	5.0	21.66	16.53	76.32	10.24	0.87	300
SSV 7073	51.72	29.83	4.85	21.80	5.1	22.84	15.87	71.36	9.37	1.28	297
SSV 20984	42.69	20.57	3.19	25.48	5.3	18.31	12.79	69.85	7.47	0.89	344
S.E. ±	8.77	7.97	0.78	6.05	0.11	1.46	1.34	3.46	1.26	0.37	28.7
CD at 5 %	26.33	23.93	2.37	18.17	0.35	4.40	4.02	10.39	3.80	1.14	86.2

Table 22. Chemical analysis of sweet sorghum juice during storage (NSSH 104)

Time (hrs)	^{0}Brix	pH	Crushed stem (g)	Juice vol. (ml)	Extraction (%)	Total sugars (%)	Reducing sugars (%)	Sucrose (%)
0	21.00	5.05	2300	1150	50.00	18.10	0.59	17.51
6	21.50	5.02	3000	1400	46.70	18.40	1.17	17.23
12	21.50	5.00	2000	900	45.00	18.40	1.57	16.83
24	21.50	4.94	2500	1100	44.00	18.35	2.09	16.26
30	21.20	4.91	1200	500	41.70	18.75	3.06	15.69
36	21.20	4.89	2500	1000	40.00	19.15	3.65	15.50
48	23.00	4.85	1800	600	33.30	20.30	5.50	14.80
Per cent decrease								-17.00

Chapter **13**

Preparation of Syrup from Sweet Sorghum Juice

After boiling and removing molasses the juice should be heated to 103 to 105 ^{0}C. At this stage the degree ^{0}brix of the boiled juice should be checked with the help of hand refractometer. The degree ^{0}brix of syrup should be upto 70%. Then stop the boiling of syrup and pour into the stainless steel, aluminium or G.I. sheets container for settlement and sedimentation of the remaining suspended particles. When syrup become cold then remove the top flouting particles on the syrup with the help of steel shiba. After cooling, add 0.04% citric acid into the syrup to get good attractive colour. Then add 50g sodium benzoate or 0.1% potassium metabisulfate in the syrup for preservation for longer time. Then fill different size (100, 250, 1000 or 5000 ml) presterilized bottles with syrup and keep 1 cm space at the top of the bottle for sealing with the help of paraffin wax. After solidification of wax, seal the bottle with the help of bottle sealing machine. Then store the syrup bottles at cold and clean place until sold in the market. From good sweet sorghum crop 4 to 5 tones per hectare syrup can be obtained.

The sweet sorghum syrup was found to be especially rich in minerals such as iron, zinc, calcium, sulphur, vitamin C, proteins, riboflavin and nicotinic acid and was free of any pesticide residues (Tables 23 and 24). The syrup in addition to being table syrup to be eaten with bread or pancakes can be used in salad dressing, preparation of various types of juices, chocolates, as a sweetener in baked goods or as ice-cream topping. It could be also provide a syrup base for pharmaceutical formulations. Nutritive constituents of sweet sorghum syrup (Madhura) and its comparison with honey are presented in Table 25 and Plate No. 10.

Table 23. Minerals and polyphenols content in sweet sorghum syrup (milligram/100 g).

Genotype	Iron	Calcium	Phosphorus	Polyphenols
RSSV 9	18.3	18.60	159.4	157.9
RSSV 47	15.6	17.77	148.1	153.6
NSS 104	17.0	16.86	114.4	175.4
Wray	14.9	17.70	137.3	146.6
Keller	18.1	18.17	160.1	163.3
SSV 84	17.0	18.93	115.6	192.4

Table 24. Chemical composition of syrup from sweet sorghum.

Parameter	Content
Total sugars (%)	86.15
Reducing sugars (%)	59.07
Calcium (%)	0.2
Magnesium (%)	0.05
Sulphur (%)	3.83
Zinc (ppm)	2.92
Copper (ppm)	0.75
Iron (ppm)	158.08
Manganese (ppm)	3.5

Table 25. Constituents of Madhura syrup and its comparison with honey.

Constituent	Madhura syrup	Honey
Total soluble solids (%)	77.00	81.00
Reducing sugars (%)	70.30	70.40
Protein (%)	1.65	-
Ash (%)	3.69	0.59
Calorific value (kcal/g)	2.60	3.26
Calcium (mg/100 g)	160.00	5.00
Phosphorus (mg/100 g)	11.00	4.10
Iron (mg/100 g)	0.86	0.59
Sodium (mg/100 g)	86.00	4.70
Potassium (mg/100 g)	1810.00	90.00
Sulphur (mg/100 g)	Not detected	8.00

Riboflavin (B_2) (mg/100 g)	10.00	0.06
Vitamin C (mg/100 g)	11.50	5.00
Nicotinic acid (mg/100 g)	153.00	32.00
Benzoic acid (mg/100 g)	Not detected	-
Added colouring matter	None	-
Pesticide residues	Not detected	-

Effect of planting and variety on the polyphenols and mineral contents in sweet sorghum syrup:

The polyphenols content of different sweet sorghum genotypes ranged from 142.2 to 198.3 mg/100g. In all the sowing dates, all the genotypes had significantly lower polyphenol content than the check SSV 84. The overall effect of sowing dates shows that Wray (146.6 mg/100g) had significantly lower polyphenols followed by RSSV 47 (143.6 mg/100g) and RSSV 9 (157.9 mg/100g). In June planting RSSV 9 (19.3 mg/100g) showed highest iron content followed by Keller (18.7 mg/100g). In December planting Keller and RSSV 9 (18.8 and 18.4 mg/100g, respectively) showed highest iron content. In April planting NSS 104 (18.1 mg/100g) and RSSV 9 (17.3 mg/100g) showed higher iron content than the rest of the genotypes (Table 26). The calcium content of sweet sorghum syrup ranged from 15.7 to 21.0 mg/100g. The genotypes Keller (18.5 mg/100g) and RSSV 9 (18.2 mg/100g) had higher calcium content at June planting. December planting RSSV 9 showed highest calcium content followed by Wray (17.99 mg/100g) and NSS 104 (17.87 mg/100g). The SSV 84 (21.0 mg/100g) showed highest calcium content followed by Keller (19.97 mg/100g) and Wray (18.47 mg/100g) in April planting. The phosphorus content in syrup ranged from 102.8 mg/100g to 170.9 mg/100g. June planting indicated that RSSV 47 (155.6 mg/100g) and Keller (153.5 mg/100g) had higher phosphorus content than the rest of the genotypes. In December planting Keller (170.9 mg/100g) recorded highest phosphorus followed by RSSV 9 (163.5 mg/100g). The genotype RSSV 9 (165.6 mg/100g) recorded highest phosphorus followed by Keller (156.1 mg/100g) at April planting. The overall effect of sowing dates showed that the syrup from Keller, RSSV 9 and RSSV 47 recorded highest phosphorus content than the rest of the genotypes studied.

Table 26. Effect of planting dates on mineral and polyphenols content of sweet sorghum syrup.

Genotype	Polyphenols, mg/100g			Iron (mg/100g)			Calcium (mg/100g)			Phosphorus (mg/100g)		
	Jun.	Dec.	Apr.	Jun.	Dec.	Apr.	Jun.	Dec.	Apr.	Jun.	Dec.	Apr.
RSSV 9	169.8	143.7	160.2	19.3	18.4	17.3	18.2	19.6	17.9	149.2	163.5	165.6
RSSV 47	143.8	156.5	160.4	15.5	16.4	15.0	17.5	16.6	16.2	155.6	145.1	143.5
NSSV 104	171.8	183.1	171.4	16.1	16.9	18.1	15.7	17.9	17.0	124.5	114.8	102.8
Wray	142.2	151.4	146.3	14.3	15.3	15.0	16.8	18.0	18.5	137.1	126.5	148.3
Keller	170.8	155.3	163.8	18.7	18.8	16.7	18.5	16.0	20.0	153.5	170.9	156.1
SSV 84	188.3	190.6	198.3	16.1	16.8	18.1	17.9	17.9	21.0	116.2	122.4	108.2
S.E. ±	1.54	2.11	2.29	0.51	0.38	0.34	0.39	0.40	0.26	2.68	2.49	2.48
C.D. at 5%	4.65	6.37	6.92	1.58	1.15	1.02	1.19	1.22	0.78	8.08	7.52	7.47

Chapter **14**

Cost of Syrup and Jaggery Preparation from Sweet Sorghum

The cost of processing of the sweet sorghum stalks for production of crude as well as table syrup is worked out by NARI Phaltan. It is contemplated that there are two types of processing channels exist for syrup manufacturing. The first channel involves the processor who procures the stalks and processes it for syrup production that again sells the syrup to another agent for further processing to produce ethanol. The second channel comprises a single agent who runs the entire process from procuring stalks to ethanol production. The cost of production of syrup from sweet sorghum stalks was worked out to be Rs. 12.10 (Table 27). However, the cost of production of the ethanol from sweet sorghum stalks directly (single agent channel). Rs. 13.11/L which could be comparable to this syrup-to-ethanol channel, after incurring the cost of processing such as filtration and dilution to produce ethanol from sweet sorghum syrup. Therefore, it could be concluded that there is no significant difference in profit structure between these two production channels.

Table 27. Economics of production of sweet sorghum syrup

Sr No	Constituent	Rs.
A.	Working capital	32,474.80
i.	Cost of cultivation upto harvesting	-
ii.	Cost of purchase of stalk @ Rs. 500/t (Rs. 500 x 30 t)	15,000.00
iii.	Cost of transport of sweet sorghum stalks from field to syrup unit @ Rs. 150 or 300/batch; Rs. 150 or 300 x 28 batches	8400.00

iv.	Cost of stripping	--
v.	Crushing cost (Rs. 222.10 x 28 batches)	6218.80
vi.	Processing cost (2 laborers x Rs. 51 x 28 batches)	2856.00
vii.	Cost of threshing grain on hire basis (Rs. 25 x 159)	-
viii.	Cost of bottling and labeling @ Rs. 3 x 4094 bottles	-
B.	Interest on working capital @ 12 % p.a. for half year	1948.49
C.	Fixed capital of syrup processing unit	1,30,500.00
D.	Interest on fixed capital @ Rs. 10 % p.a. for half year	6525.00
E.	Maintenance cost on fixed assets @ Rs. 0.5 % p.a. for half year	326.25
F.	Additional returns from grain @ Rs. 600 x 15 g	-
G.	Total cost of syrup production	41,274.54
H.	Quantity of syrup obtained (kg)	3412.00
I.	Cost of syrup per kg (Rs/kg)	12.10

Source: NARI, Phaltan, Dist: Satara, Maharashtra.

Economics of production of syrup and jaggery is given in Table 28. It shows that if SSV 84 is cultivated for syrup or jaggery the net profit of Rs. 41000/- from syrup and Rs. 15,750/- from jaggery per hectare can be accrued in a period of 130 to 150 days.

Table 28. Economics of production of syrup and jaggery from sweet sorghum.

Sr. No	**Particular**	**Syrup (Rs.)**	**Jaggery (Rs.)**
A.	**Cost of cultivation of sweet sorghum**	7,000.00	7,000.00
	1. Yield of green stalk (Shraded) 50 t/ha		
	2. Yield of grains/ha, 15 q/ha		
	3. Yield of juice from 50 tonnes, 20 tonnes		
	4. Yield of syrup/ha from 50 tonnes, 4 tonnes		
	5. Yield of jaggery, 3.5 tonnes		
B.	1. Cost of production of syrup/jaggery (Inclusive of labour charges, fuel charges etc.)	13,000.00	13,000.00
	2. Cost of bottling of syrup (inclusive cost of bottles, preservatives), each bottle of 450 g (approx. 900 bottles)	53,000.00	--
	Total cost	**73,000.00**	**20,000.00**

C.	1. Income from syrup (Rs. 12/bottle and jaggery Rs. 8.5/kg)	1,08,000.00	29,750.00
	2. Income from grain (Rs. 400/q for 15 q/ha)	6,000.00	6,000.00
	Total income	1,14,000.00	35,750.00
D.	**Net profit**	**41,000.00**	**15,750.00**

Chapter 15

Preparation of Jaggery from Sweet Sorghum Juice

The mixture of syrup to be continued to heat to get the 110 to 112 ^{0}C temperature of the mixture. While heating the syrup add 200 ml edible oil for 750 liter juice. This will help to increase the syrup temperature and avoid sticking to the container or boiling pan. If you don't have thermometer to see the syrup temperature, then we can see the appropriate stage of jaggery by putting small quantity of syrup in to cold water and test the hard ball. If that syrup ball were comes hard then remove the boiling pan from the fire and pour the jaggery syrup in to another pan or cement box for cooling. At the same stir it continuous to avoid crystallization. After getting 70 to 75^{0}C temperature fill the empty buckets already inside covered with clean cloth and varying various capacitiesfrom 1 to 30 kg. After completely cooling, remove the bucket cloth and rap that jaggery depth into ploythene bag or gunny bag and store into dry and clean place until further use. Approximately 3.5 to 4.0 tones jaggery can be obtained from one hectare of sweet sorghum (Plate No. 11).

Juice quality parameters for good quality syrup and jaggery preparation:

- The juice ^{0}brix should be higher than 21.
- Sucrose percentage should be higher than 19.
- Glucose percentage should be less than 0.5.
- Soluble nitrogenous compounds should be higher than 32 mg/100ml of juice.
- Phosphate pentaoxide should be higher than 30 mg/100ml

of juice.

- Nitrogen phosphorus ratio should be less than 1.5.
- pH of the juice should be close to the 5.0.
- Starch and Protein content in the juice should be very minute.
- Suspended particles in the juice should be less.
- Chlorophyll and phenolics compounds should be removed completely.
- Juice should be extracted from freshly harvested canes (to avoid stailing effect).

Sweet Sorghum varieties for syrup, jaggery and ethanol production:

The juice quality of sweet sorghum is depending upon the variety of the sweet sorghum. The quality of syrup, jaggery and ethanol mostly depends upon the quality of juice. For getting good quality juice and from that good quality of final product, it is necessary to use the recommended variety.

Recommended varieties:

SSV 74, SSV 84, SSV 108, SSV 225, SSV 7073, NSS 04, NSS 104, NSS 208, Warey, kellar, B.J 248, RSSV 09, RSSV 47 and Madhura. Characteristics of some selected sweet sorghum genotypes are given in Tables 29 and 30.

Table 29. Comparative study of some sweet sorghum varieties

Variety	TSS (^{0}Brix)	Stalk yield (t/ha)
RSSV 9	18.87	40.8
RSSV 12	18.41	60.2
RSSV 21	20.56	63.2
RSSV 26	19.10	54.3
RSSV 28	20.84	52.6
RSSV 30	19.38	67.2
RSSV 84	18.38	39.1
SSV 84	19.00	32.5
CSV 19SS	19.50	37.5
CSH 22SS	20.00	42.5

Table 30. Comparative characteristics of some selected sweet sorghum genotypes.

Character	SSV 84	CSV 19SS	CSH 22SS
Evolution of genotype	Selection from IS 23568	RSV 2 x SPV 462	ICSA 38A x SSV 84
Days for 50% flowering (days)	80-85	76-80	82-85
Maturity period (days)	120-125	115-120	120-125
Height of plant (cm)	210-230	240-270	240-270
Cane yield (t/ha)	30-35	35-40	40-45
^{0}Brix	18-20	18-20	19-20
Reducing sugars (%)	1.0-1.4	0.9-1.2	1.0-1.5
Non-reducing sugars (%)	13-15	14-16	11-12
Total sugars (%)	14.0-16.4	15-17	12-14
Grain yield (q/ha)	10-12	10-13	10-15
Juice yield (L/ha)	8000-10000	8500-11000	12000-14000
Ethanol (%)	9-10	8-10	8.5-9.0
Ethanol yield (L/ha)	900-1000	900-1200	1000-1200
Syrup yield (t/ha)	3.5-4.5	3.5-4.5	3.5-4.5
Jaggery yield (t/ha)	3.0-4.0	3.0-4.0	3.0-4.0

During 1992-93 *Kharif* season twenty four sweet sorghum entries in ASSVT and ISSVT were tested for yield and quality of juice (Table 31). From these entries ten promising entries were used for the preparation of syrup and jaggery and their quality was studied (Table 32). The entries GSSV 148, RSSV 1, RSSV 2, RSSV 4, SSV 84, GSSV 206, AKSS 4, AKSS 10, and PBS 2 were found to be the most promising entries for the preparation of syrup and jaggery.

Three years pooled data of green stalk weight, grain yield, and quality traits are presented in Table 33, which indicated that the sweet sorghum variety SSV 84 had higher green stalk yield (39.35 t/ha) over other genotypes studied. The grain yield ranged from 12.75 to 19.65 q/ha. The juice quality studied revealed that the genotype SSV 84 recorded highest TSS (18.94%). Maximum non-reducing sugars (12.87%) with the lowest reducing sugars (1.81%) were observed in SSV 53. The

Table 31. Yield and juice quality of sweet sorghum varieties tested in advanced sorghum varietal trials.

Genotype	Green stalk yield (t/ha)	Grain yield (q/ha)	pH	Total soluble solids (^{0}Brix)	Reducing sugars (%)	Non-reducing sugars (%)	Total sugars (%)	Commercial cane sugars (%)	Commercial cane sugars (q/ha)	Starch (%)
GSSV 148	27.25	7.37	4.7	21	3.25	13.18	16.43	7.44	2.03	0.744
AKSS 5	37.80	8.40	4.8	20	2.94	11.89	14.83	6.19	2.34	1.084
RSSV 2	27.80	13.03	4.7	22	2.17	13.78	15.95	7.76	2.16	0.768
RSSV 4	32.30	8.59	4.7	22	3.01	14.87	17.88	8.89	2.87	0.744
NSS 1	39.20	14.40	4.7	19	3.96	10.21	14.17	4.95	1.94	0.839
SSV 84	41.10	18.86	4.7	22	3.73	13.18	16.91	7.14	2.93	0.768
S.E. ±	1.40	2.20	0.03	1.15	0.58	1.47	1.24	1.23	0.38	0.120
CD at 5 %	4.80	6.53	0.13	3.46	1.74	4.43	3.73	3.71	1.17	0.361

Table 32. Yield and juice quality of sweet sorghum varieties tested in advanced sorghum varietal trials.

Genotype	Green stalk yield (t/ha)	Grain yield (q/ha)	pH	Total soluble solids (^{0}Brix)	Reducing sugars (%)	Non-reducing sugars (%)	Total sugars (%)	Commercial cane sugars (%)	Commercial cane sugars (q/ha)	Starch (%)
GSSV 151	38.2	26.12	4.7	18.5	3.85	11.57	15.42	6.51	2.49	0.635
GSSV 186	24.5	43.65	4.7	16.5	3.57	10.58	14.15	6.08	1.49	0.594
GSSV 195	29.5	22.49	4.8	22.0	2.45	9.79	12.24	3.63	1.06	0.916
GSSV 206	25.5	15.22	4.7	18.5	3.78	12.04	15.82	7.00	1.79	0.768
AKSS 4	38.4	19.18	4.7	21.5	1.28	12.76	14.04	6.86	2.63	1.175
AKSS 10	27.3	6.28	4.7	20.5	2.94	12.83	15.77	7.22	1.97	1.025
PBS 2	32.6	23.81	4.7	19.0	2.72	13.25	15.97	8.10	2.64	1.045
RSSV 5	47.4	18.52	4.8	19.0	4.38	10.15	14.53	4.89	2.32	0.792
SSV 84	52.5	16.54	4.7	22.0	2.87	12.90	15.77	6.85	3.60	0.768
S.E. ±	2.00	2.20	0.04	1.78	0.87	1.23	1.16	1.26	0.70	0.184
CD at 5 %	10.00	6.71	0.12	5.34	2.62	3.69	3.50	3.79	2.11	0.554

Table 33. Syrup and jaggery quality of promising genotypes of sweet sorghum.

Genotype	Syrup						Jaggery				
	TSS (^{0}Brix)	pH (6.5 % solution)	Acidity (%)	Reducing sugars (%)	Non-Reducing sugars (%)	Total sugars (%)	pH (6.5 % solution)	Acidity (%)	Reducing sugars (%)	Non-Reducing sugars (%)	Total sugars (%)
GSSV 148	61.0	4.9	2.41	7.4	45.2	52.6	5.0	3.26	11.7	48.6	60.3
RSSV 1	63.5	4.9	2.81	7.5	40.9	48.5	5.0	3.66	12.0	48.3	60.3
RSSV 2	60.5	5.0	2.41	6.8	40.6	47.4	5.0	4.28	8.6	55.2	63.8
RSSV 4	58.5	4.9	2.37	6.8	39.2	46.0	4.9	3.84	13.3	50.8	64.1
SSV 84	59.0	4.9	2.50	5.5	44.4	49.9	5.0	3.35	11.9	53.4	65.3
GSSV 151	60.0	4.9	2.90	5.6	44.9	50.5	4.9	4.24	7.3	60.7	68.0
GSSV 206	59.0	4.9	2.86	9.3	38.1	47.4	5.0	4.28	13.5	45.0	58.5
AKSS 4	59.0	5.0	2.14	6.4	37.0	43.4	5.0	4.32	9.4	47.3	56.7
AKSS 10	61.0	5.0	2.90	6.3	41.3	47.6	5.0	4.13	8.8	48.7	57.5
PBS 2	66.0	4.8	2.72	6.6	40.8	47.4	4.8	4.30	9.6	52.9	62.5
S.E. $\pm$	2.23	0.06	0.25	1.03	2.67	2.40	0.06	0.39	2.02	4.34	3.46
CD at 5 %	6.72	0.18	0.77	3.10	8.04	7.20	0.20	1.17	6.08	13.05	10.40

Table 34. Physico-chemical properties of sweet sorghum and their products (3 years pooled data).

Genotype	Green stalk (t/ha)	Grain yield (q/ha)	Juice					Syrup				Jaggery			
			TSS (°Brix)	pH	RS (%)	NRS (%)	TS (%)	pH	RS (%)	NRS (%)	TS (%)	pH	RS (%)	NRS (%)	TS (%)
SSV 84	39.35	12.75	18.94	4.60	2.92	10.80	13.72	5.12	13.59	34.82	48.41	5.36	14.95	37.34	72.29
SSV 53	31.90	12.94	17.36	4.79	1.81	12.87	14.68	5.13	12.36	39.09	51.45	5.25	14.47	58.44	72.91
SSV 119	34.32	14.65	16.92	4.68	3.18	10.35	13.53	5.12	6.70	43.28	49.98	5.30	10.40	57.79	68.19
HES 4	27.35	19.65	15.88	4.82	2.44	11.00	13.44	5.13	9.00	13.50	40.50	5.34	11.00	56.00	67.00
S.E. ±	1.28	1.26	1.10	0.08	0.52	0.96	0.49	0.005	2.72	11.47	4.23	0.04	2.02	8.73	2.54
CD at 5 %	4.05	3.65	3.30	0.26	1.56	2.88	1.48	0.015	8.17	34.40	12.69	0.13	6.07	26.21	4.64

RS = Reducing sugars, NRS = Non-reducing sugars, TS = Total sugars,
pH = 6.5% solution of syrup and jaggery used.

reducing and non-reducing sugars content in syrup ranged from 6.70 to 13.59% and from 34.82 to 43.28% respectively (Table 34). Maximum total sugars in syrup (51.45%) were observed in SSV 53. The genotype SSV 84 showed 57.34% non-reducing sugars and 14.95% reducing sugars in jaggery that indicated its superiority.

Jaggery prepared from the juice of sweet sorghum and sugarcane at the same time and then analyzed for their various characteristics (Table 35). Considering all the above factors which determine the quality of jaggery it was concluded that the jaggery prepared from the juice of sweet sorghum is more or less as good as that prepared from the juice of sugarcane.

Table 35. Chemical composition of jaggery prepared from sweet sorghum and sugarcane juice.

Constituent	Sweet sorghum jaggery	Sugarcane jaggery
Moisture (%)	6.52	6.24
Hardness	3.10	3.30
Sucrose (%)	78.10	84.20
Reducing sugars (%)	8.80	7.50
Colour in solution (6.5 % solution)	378	305
Crystal (%)	58.10	64.20
Sulphate ash (%)	7.60	4.30
pH (6.5% solution)	5.50	5.60

Chapter 16

White Sugar Production from Sweet Sorghum

Sugar production from sweet sorghum on pilot scale is given in following flow chart (Figure 2) and their some important points to be consider while preparation of white sugar are mentioned herewith for further study.

Following observations must be considered while white sugar production from sweet sorghum:

1. Juice extraction percentage of sweet sorghum is upto 38 - 40% (low extraction rate).
2. Fiber percentage of sweet sorghum is upto 36-38% (higher level).
3. Purity of sweet sorghum juice is very low.
   a. Purity is varying between 64.84 to 71.36%.
   b. CCS percentage is varying between 7.76 to 9.32%.

i. From the sugar factories reports, it is seen that the juice extraction percentage and juice purity both being much on a lower side. These parameters indicate that sugar content in sweet sorghum will be less thereby the sugar recovery percentage will be much below the economical level as compared to sugarcane.

ii. Higher percentage of fiber in the juice will cause problems at the mills and higher imbibitions rates will have to be resorted too.

iii. It is also feared that higher starch content is a major constituent of the juice will pose problems in the process, such as viscosity. The rate of crystallization will also slow down tremendously.

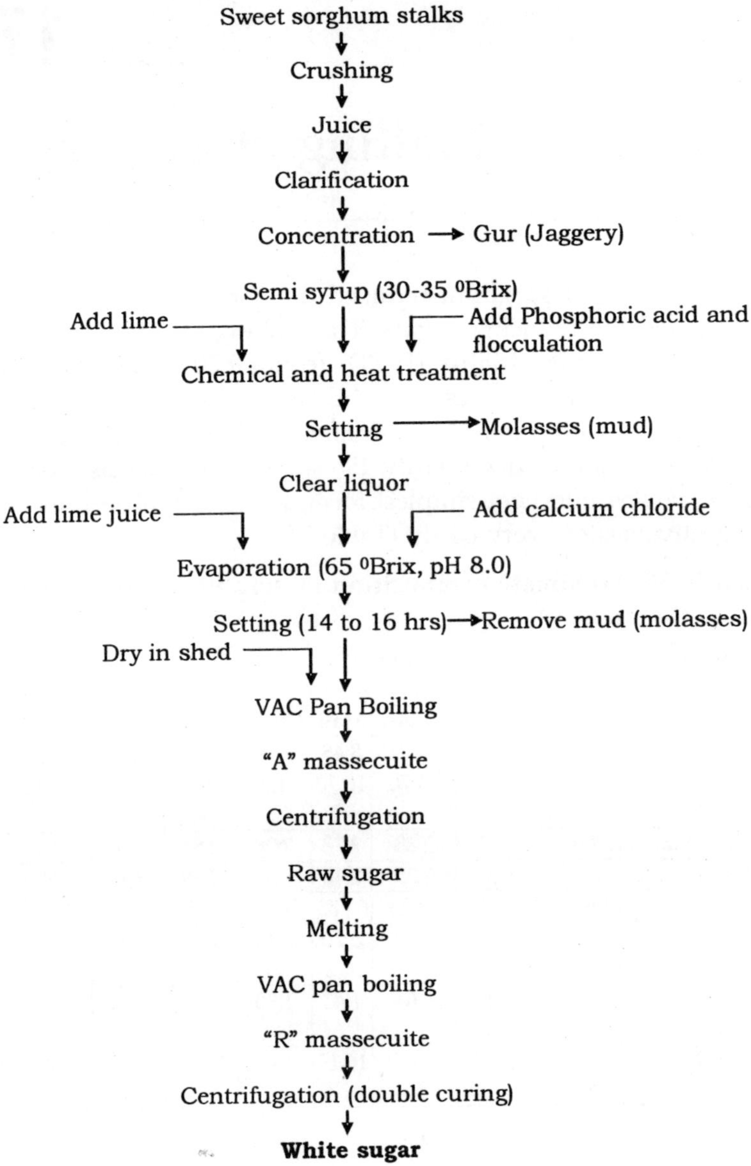

Figure 2. Preparation of white sugar from sweet sorghum

Chapter 17

Malting of Sorghum

The malting of sorghum reduced the content of starch and increased the reducing sugars. The malting also improved *in vitro* protein digestibility (IVPD) from 80.30 to 83.60%. The malted cereals and pulses are mostly utilized in the preparation of weaning food (infant food). Malted cereals as well as pulses make all major nutrients easily digestible by hydrolysis process and convert into very simplest form, which can absorb in the digestive system very easily (Table 36).

Table 36. Proximate composition of sorghum and sorghum malt.

Constituent (%)	M 35-1		Swati		CSH 5		PVK-801	
	Grain	Malt	Grain	Malt	Grain	Malt	Grain	Malt
Moisture	10.13	8.13	11.02	8.48	10.83	8.93	10.00	7.91
Crude protein	11.60	11.60	10.70	10.70	11.30	11.30	9.45	9.42
Crude fat	2.93	2.87	2.83	2.78	2.98	2.81	3.00	2.90
Crude fibre	2.13	2.10	2.35	2.10	2.48	2.39	2.10	2.00
Starch	76.23	68.13	77.24	67.83	76.81	66.35	74.20	59.50
Reducing sugar	0.20	1.40	0.10	0.70	0.10	1.40	0.30	7.00
Total minerals	1.70	1.40	1.61	1.43	1.30	1.10	1.20	0.90
Malting losses	-	13.67	-	16.15	-	17.50	-	19.00

Chapter **18**

Use of Sweet Sorghum in Brewing Industry

In sole sorghum beer (without adjunct), the bitterness and free amino nitrogen content were low compared to sorghum adjunct beer, which had highest free amino nitrogen content and essential for the beer flavour (Table 37).

Brewing is said to raise the nutritional value of sorghum since it adds Vitamins, neutralizes most of the tannins, hydrolyzes the starch to more digestible forms and increases the availability of minerals and vitamins. South African studies indicate that iron is 12 times more available in sorghum beer than in boiled sorghum gruel, riboflavin may be almost twice and thiamine almost a third more available while niacin's availability remains unchanged (Figure 3).

Enzymes in the malt begin to saccharify the starch. Bacteria then act on a part of the sugar to form lactic acid and some B vitamins. This sweet wort may be drunk after less than one day's fermentation. It contains more B vitamins and negligible amount alcohol. After the first day the mixture is heated and various fungal spores germinate and lead to the production of hydrolytic enzymes. A little millet or sorghum may be specially germinated, then dried as a yeast source. The yeasts multiply and fermentation develops, with the production of alcohol, more B vitamins and new proteins, while the bacteria continue to develop lactic acid. The brew may be drunk after 4-5 days, and in some areas more malted grain may be added to it a day before it is used.

Sorghum grains

↓

Soaking

↓

Germination

↓

Drying of malted grains

↓

Ground to fine flour

↓

Malted flour

↓

Add cold water to prepare slurry

↓

Add more pounded/steeped sorghum grains in it

↓

Keep for fermentation

↓

Filtration

↓

Bottling of malt beer

↓

Storage

Figure 3. Flowchart for preparation of sorghum beer.

Table 37. Chemical composition of beer from sweet sorghum

Constituent	Malt beer	Adjunct beer
Protein (%)	2.68	1.98
Free amino nitrogen (%)	74.61	119.01
Total sugars (%)	1.54	0.95
Reducing sugars (%)	0.05	0.02
Alcohol (%)	4.66	4.50
Colour (EBC)	5.89	8.58
Specific gravity	1.03	0.98
Bitterness (BU)	6.56	7.58

Chapter **19**

Starch and Maltodextrin from Sorghum

The starch prepared from black sorghum grain have a vital role and potential raw material for the production of maltodextrin and it may be explored for industrial or commercial exploitation in different food products as fat replacer to reduce the calorific value. Metabolisms of maltodextrins are similar to starch. Food products for diabetic patients can be prepared by using maltodextrins as a fat replacer. Maltodextrins gives hypoglycemic activity and is also a promising food substance to heart patients because it acts as a very good hypocholesterlemic agent.

The maximum amount of maltodextrin was liberated at 4.5% concentration of hydrochloric acid hydrolysis process at 120 minutes interval i.e., 95 per cent and 92 per cent for PVK 801 and CSH 9 sorghum varieties and their dextrose equivalents were 9 and 8 respectively. The maltodextrin contained 98 and 97 per cent total carbohydrate in PVK 801 and CSH 9 respectively (Table 38). The results of reheological attributes particularly the pasting behaviour of maltodextrinrevealed that the viscosity at 40^{0}C was found to be maximum and then decreased as the temperature increased. The microscopic examination of maltodextrin granules revealed that they are mostly irregular shaped particle aggregates with 4 – 5 μ in size (Table 39). The results indicated that good quality cookies and toffees could be prepared using maltodextrin with 3 per cent replacement of shortening. Maltodextrins are metabolized similar to starch. Thus, fat replacers like maltodextrin can be recommended for use in food products for diabetics, it exhibits hypoglycemic activity. Moreover, it will be a very promising for heart patients also as it acts as a very good hypocholesterlemic agent. The comparative characteristics of starch and maltodextrin of sorghum is given in Table 40 and Figure 4.

Table 38. Chemical composition and physical parameters of sorghum starch.

Parameter	Content (%)	
	PVK-801	CSH-9
Moisture	9.80	9.50
Protein	1.10	1.29
Fat	0.90	0.95
Ash	0.24	0.45
Crude fiber	0.73	0.67
Total carbohydrate	96.81	95.60
Yield of starch	75.50	72.85
Granule shape	Polygonal to round	Polygonal to round
Granule size	8.0-9.5 μ	8.0-9.1 μ
Colour of starch	Bright white	Mild white
Iodine affinity	4.4	3.9
Bulk density	0.59	0.58
Molecular weight		
Amylose (daltan)	2,45,000 $\pm$ 10,000	2,45,000 $\pm$ 10,000
Amylopectin (daltan)	> 0.59 x 10^6	> 0.59 x 10^6

Table 39. Physico-chemical constituents of maltodextrin.

Parameter	Content (%)	
	PVK-801	CSH-9
Moisture	8.0	7.6
Protein	0.05	0.07
Carbohydrate	98.0	97.0
Fat	0.07	0.08
Ash	0.6	0.7
Bulk density	0.579	0.575
Granule shape	Irregular	Irregular

Granule size	3-5 μ	4-5 μ
Granule colour	Bright white	White
Molecular weight		
Amylose (daltan)	1,75,000 $\pm$ 5000	1,75,000 $\pm$ 5000
Amylopectin (daltan)	> 0.59 x 10^6	> 0.59 x 10^6
Dextrose equivalent	9	8

Table 40. Physico-chemical properties of sorghum starch and maltodextrin.

Constituent	Starch	Maltodextrin
Moisture (%)	9.50-9.80	7.5-8.0
Protein (%)	1.10-1.29	0.05-0.07
Fat (%)	0.90-0.95	0.07-0.08
Ash (%)	0.24-0.45	0.6-0.7
Fiber (%)	0.67-0.73	0.4-0.5
Total carbohydrates (%)	95.60-96.81	97-98
Yield (%)	73-76	-
Granule shape	Round, polygonal	Irregular
Granule size (μ)	8.0-9.5	3.0-5.0
Colour of granule	Bright white	Mild white
Iodine affinity	3.9-4.4	-
Bulk density	0.58-0.59	0.57-0.59
Molecular weight of Amylose (daltan)	2,45,000	1,75,000
Molecular weight of Amylopectin (daltan)	> 0.59 x 10^6	> 0.59 x 10^6

Sorghum grains

↓

Cleaning (Removal of foreign matters)

↓

Steeping in water (20 hrs, at 50 ^{0}C and 4.5 pH)
(750 ppm KMS, 0.002 M $HgCl_2$, few drops of toluene in water)

↓

Washing with clean water

↓

Blending / preparation of slurry (1:3 w/v, seed : water)

↓

Separation of starch
(Squeezing of slurry through muslin cloth and then passing through 150 μ sieve)

↓

Protein denaturation
(Add 1.5 g NaCl and 5 ml Toluene in slurry)

↓

Shaking for 1 hr

↓

Centrifugation at 4200 RPM for 15 min.

↓

Isolation of starch

↓

Drying (Vacuum drying)

↓

Air drying for overnight

↓

Starch

↓

Acid hydrolysis
(4.5 % HCl conc. 50 ^{0}C temp. and 120 min. time)

↓

Isolation —Vacuum drying→ **Maltodextrin**

Figure 4. Flow diagram for the isolation and purification of starch and maltodextrin.

Chapter **20**

Production of High Fructose Syrup from Sorghum Starch

The sorghum genotypes cultivated in *kharif* season contain about 68 to 76 per cent starch. The starch is a potential raw material being used in several cloth, paper, dye as well as in food industry in India and foreign country also. Therefore, the production of starch from *kharif* sorghum is one of the most prominent sources because most of the high yielding hybrids mostly cultivated during *kharif* season. For the preparation of value addition products from rain damaged *kharif* season sorghum and also to exploit it for industrial utilization, preparation of starch and its conversion into other valuable products such as high fructose syrup, maltodextrin, sweeteners, carboxymethylated starch, glucose, and sorbitol is most essential.

Fructose is the sweetest sugar among the natural sugars. The relative sweetness of fructose is a function of concentration, temperature and pH, with the maximum sweetness value being found in cooled, dilute solutions that is neutral or slightly acidic. Fructose has a short sweetness duration that tends to enhance the aroma and flavours of fruits. The high fructose syrup can be prepared from sorghum (Figure 5a and 5b) as well as sweetener too (Figure 5c). The refractive index for high fructose syrup (HFS) prepared from CSH 18 starch (1.453) was same as that of HFS from glucose sample (1.455). Colour of the syrup prepared from starch was yellowish and the colour of the HFS prepared from control glucose sample was whitish yellow (Table 41).

Commercially sucrose is usually marketed as sucrose or medium invert sugar, which is liquid and contains 50 parts sucrose, 25 parts dextrose and 25 parts fructose. High fructose sorghum syrup is similar to total sucrose invert; therefore, it replaced sucrose on weight-to-weight basis. Due to the refining to remove non-sweetener components and because of excessive microbiological control, HFS is a very high quality liquid sweetener. The predominant user of HFS is the soft drink industry. The mango squash prepared by using HFSS had recorded higher preference for all the attributes than that of control. The high fructose sorghum syrup can be effectively used to replace sucrose in the products like fruit squashes.

Sorghum starch

↓ ← Dextrinization with acid hydrolysis
← Liquification by α-amylase
← Saccharification with Glucoamylase

Glucose syrup (95-96 %) Dextrose

↓ ← Refining
← Demineralization
← Decolourisation

Purification

↓

Purified glucose and syrup

↓

Isomerization

↓ ← With glucose isomerase
Temperature 60 ^{0}C
pH 6.8-7.0

High fructose sorghum syrup (HFSS)

↓

Purification

↓ ← Decolourization using carbon celite column

HFSS (45% D.S.)

↓

Evaporation

↓

HFSS (54-55% D.S.)

Figure 5a. Production of high fructose syrup from sorghum starch

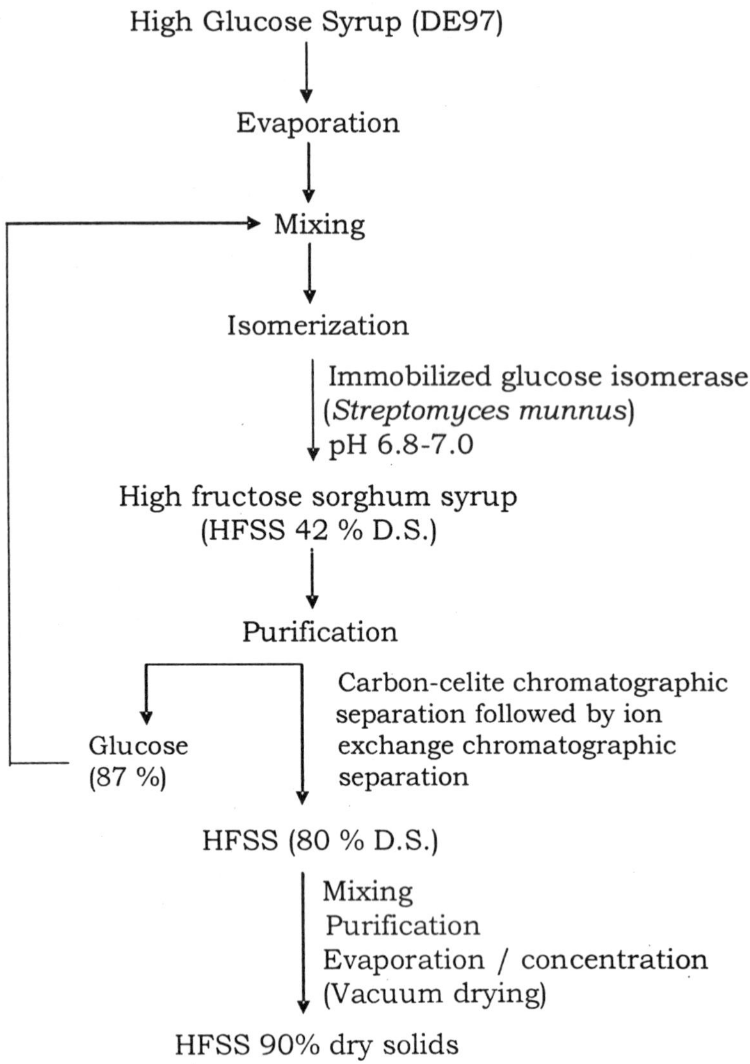

Figure 5b. Production of high fructose syrup using immobilized glucoisomerase (*Streptomyces munnus*) at optimum experimental conditions

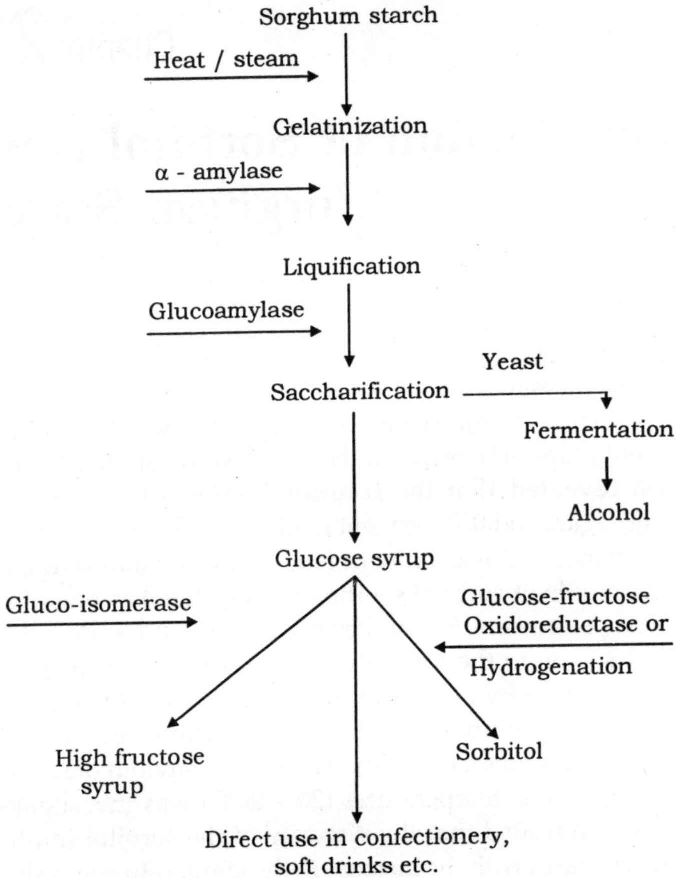

Figure 5c. Sweetener production from starch

Table 41. Physico-chemical properties of high fructose syrup.

Parameter	HFSS from CSH-18	HFS control (glucose)
Refractive index	1.453	1.455
Colour	Yellowish	Whitish yellow
Dry solids (%)	67.93	70.00
Moisture (%)	32.06	30.00
Ash (%)	0.08	0.05
Acidity (%)	0.11	0.09
^{0}Brix (TSS, %)	58.33	60.00

Chapter **21**

Production of Sorbitol from Sorghum Starch

Sorbitol solution appears to be a clear, colourless and syrup liquid with pleasantly sweet taste. The solution was essentially odourless. The sorbitol content of the solution was found to be 66 per cent (Table 42). Study of the composition of the prepared solution revealed that the solution contents 0.21 per cent reducing sugars and 0.70 per cent total sugars. In order to assess the importance and feasibility in food and other industries, the viscosity profile of sorbitol solution is considered as one of the important parameters. Sorbitol being a monosaccharide polyol, contributes little to the viscosity of the solution to which it is added. The viscosity of sorbitol solution over a wide range of shear rates and temperature has been carried out by using Haake's Rotoviscometer (RV 20). The flow behaviour of sorbitol solution at various temperatures (20 - 26°C) was investigated. The results indicated that the viscosity of the sorbitol solution linearly decreases with the increase in the temperature and shear rate (Table 43).

Various functional properties of sorbitol were extensively studied. The refractive index of the solution (at 20°C) was found to be 1.45. The relative sweetness of the 10 per cent solution was compared with the sweetness of standard sucrose. The relative sweetness of the solution was found to be 60:100. The data on the cooling effect of the solution revealed that the solution gives cooling effect within six seconds of putting it in mouth. The solution was found to be thermally stable up to 160 - 180°C (Table 44). The flash point of the sorbitol solution was found to be 300°C and the specific gravity of the solution was 1.28 at 25°C. The non-carcinogenic properties of sorbitol are making it increasing accepted sweetener in the insulin-independent mechanism of sorbitol make it an ideal sweetener for the diabetes. Sorbitol produced from sorghum starch hydrolyzes by chemical hydrogenation in presence of nickel catalyst and was used as sweetener. The milk chocolate prepared by using sucrose had scored higher value for all the quality attributes than the milk chocolates prepared by using sorbitol and sucrose in combination. Milk chocolate manufactured from sorbitol resulted in low-viscosity chocolate base, which was corrected by using additives (starch or guar gum 2 per cent).

Table 42. Biochemical properties of sorbitol solution

Parameter	Observations
Appearance	Clear, colourless, syrup liquid
Colour	Watery white
Taste	Pleasantly sweet
Odour	Essentially odourless
Assay (%) D-sorbitol	66.0
Reducing sugars (%)	0.21
Total sugars (%)	0.70
Residues on ignition (%)	0.10
Water (%)	29.5
pH (10 % solution)	6.5

Table 43. Viscosity profile for sorbitol syrup.

Temperature (^{0}C)	Apparent viscosity (Cps) Shear rate (S^1)			
	4.51	**34.73**	**97.41**	**451**
20	160	147	132	98
30	110	105	88	71
40	80	72	66	49
50	65	58	48	34
60	55	47	32	20

Table 44. Functional properties of sorbitol solution.

Parameter	Observations
Refractive index (20 ^{0}C)	1.45
Relative sweetness (10 % solution)	60:100
Cooling effect	In 6 seconds
Thermal stability	160-180 ^{0}C
Flash point	Above 300 ^{0}C
Specific gravity at 25 ^{0}C	1.28

Chapter **22**

Ethanol Production from Sweet Sorghum Juice

The sweet sorghum must be harvested at fully mature stage when the ^{0}Brix of juice is more than 19 per cent. The harvested canes are crushed with the help of three-roller horizontal cane crusher and juice is obtained. The juice must be supplemented with diammonium hydrogen phosphate (0.05g/100ml) and potassium metabisulphate equivalent to 100 ppm SO_2. For the preparation of inoculums take small quantity of juice and yeast and make a sufficient quantity of inoculums for large quantity of juice fermentation. Then the must after pasteurization at 82-83^{0}C for 30 min inoculate with old yeast inoculums and keep for the fermentation for 48 to 72 hrs and after that check fermentation is completed or not. After completion of fermentation add bentonite in it and then filter/centrifuge and centrifuged liquid then distill for separation of alcohol and get distilled alcohol, check its purity. After redistillation pure alcohol can get and that can be used as a biofuel for different purposes (Ethanol industry Plate No. 12 and 13). The detail flow chart for preparation of ethanol is given in Figures 6a and 6b.

Pilot scale Evaluation of Ethanol from the Stalk Juice of Sweet Sorghum

A pilot study in collaboration with Sri Renuka sugar factory (Manoli, Belgaum district) was conducted for the ethanol production (95%) from sweet sorghum juice. Two varieties of sweet sorghum (SSV 74 and SSV 84) and one hybrid (Madhura) were grown in 600 acres in and around the factory area. SSV 84 realized 25 and 2 tons per hectare of green cane and grain yields

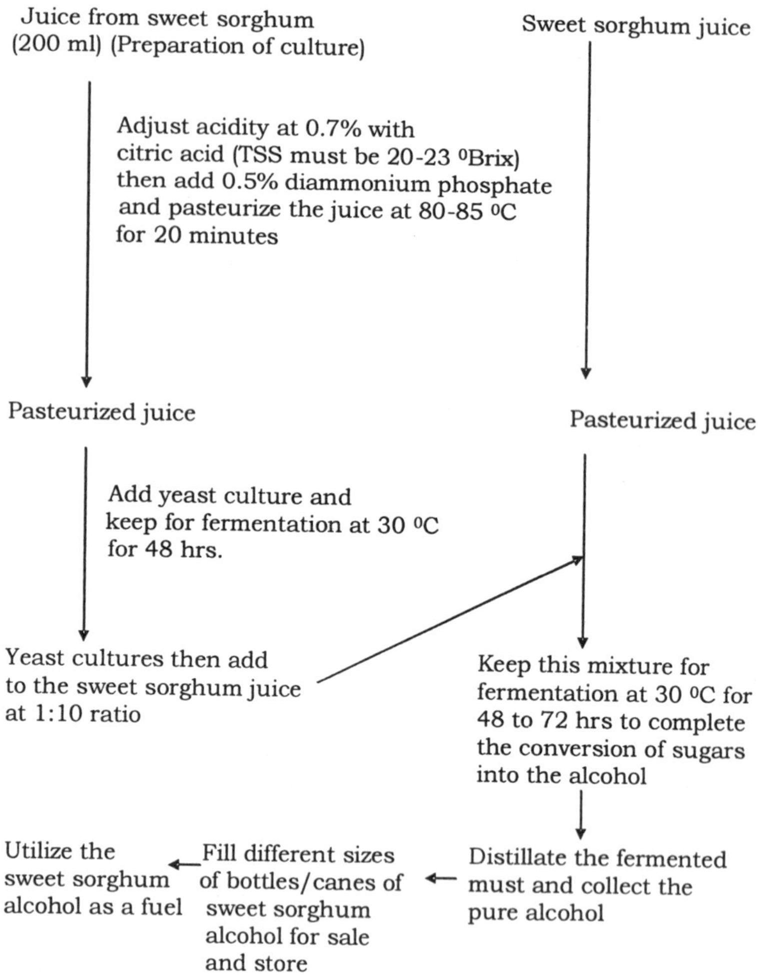

Figure 6a. Flow chart for preparation of alcohol from sweet sorghum juice in the laboratory

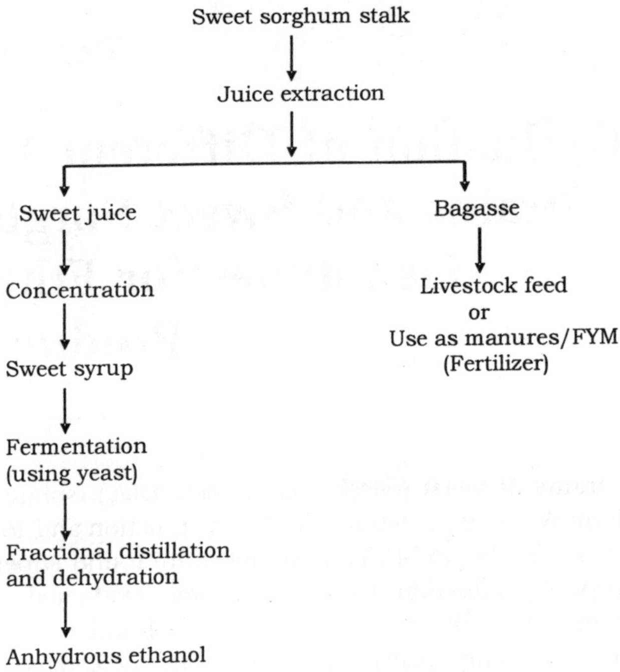

Figure 6b. Ethanol production from sweet sorghum

respectively. The sweet sorghum cane was crushed along with leaves to minimize labour; but obstructional the blades resulted thus in reducing its efficiency and increase fibre content in the juice. Hence, the cane was manually fed to the crusher. ^{0}Brix content was initially 18%, which declined to 12% after fermentation. About 112 tonnes of cane was used for fermentation using yeast (*saccharomyces cerevisiae*) in one fermentor of capacity 60 Kl over a period of 48 h. Ethanol recovery was approximately 9% of the juice, and with 46.4% bagasse obtained with 2.6% sugars. Bagasse was successfully used for the cogeneration of electricity. While production of ethanol from sweet sorghum juice the by-products obtained are aldehydes and esters and their recovery was similar to sugarcane. Further the fiber content of the juice was more when crushed with the leaves. The leftover sludge still contained 6.4% ^{0}brix, which can be attributed to the low efficiency of the yeast strain used.

Chapter 23

Evaluation of Different Yeast Strains and Sweet Sorghum Genotypes for Ethanol Production

---

Ten strains of yeast (*Saccharomyces cerevisiae*) isolated from sorghum were evaluated for alcohol production and tolerance to alcohol (Plate No.14). Four strains were found superior for the alcohol production. These strains were evaluated for their performance in the production of alcohol and fermentation efficiency in comparison to the strain (*Schizo-sacchormyces pombe*) obtained from distillers. The sugar composition of Madhura, BJ 248 and RSSV 9 realized high alcohol percentage irrespective of any yeast strain.

Chapter **24**

Economics of Ethanol Production from Sweet Sorghum

The primary characteristics determining the production of ethanol from sweet sorghum and sugarcane are presented in Table 45. The data are derived from fermentation tests carried out at Triangle Ltd's Laboratories on sweet sorghum stem samples before a full scale-crushing test. This table calculates potential ethanol yields for two scenarios: (1) where all the sugars are used for ethanol production and (2) crystalline sugar is extracted first and the residual molasses (molasses "C") is used for ethanol production. It is based on the results of two full scale crushing trials carried out at Triangle Ltd's mill at the start of the 1999 and 2000 sugarcane harvesting seasons using both the 66′ mill tandem and the diffuser lines to separate the juice from the fiber.

Table 45. Potential ethanol production from sweet sorghum (% fresh weight stems).

Variety	°Brix	Pol[1]	Sucrose purity[1]	Total Fermentable sugar (%)[1]	Ehanol[2] (%)	Ehanol[3] (L/ha)
1. Ethanol only						
Keller	17.4	12.1	69.7	13.1	4.6	2677
Cowley	18.5	12.8	69.2	14.7	5.1	2993
Is 19674	13.7	6.6	48.2	9.3	3.3	1904
Monoriedes	11.0	6.3	57.3	7.7	2.7	2057
Sugarcane	16.8	14.1	83.6	14.6	5.1	7458
2. Ethanol + crystalline sugar						
Sorghum (Keller)	-	1.1	-	2.1	0.7	561
Sugarcane	-	1.3	-	1.8	0.6	936

Source: Woods 2000.

1. ^{0}Brix = Total dissolvable solids;

   Pol = Polarity (measure of sucrose)

   Sucrose purity = % ^{0}Brix which is sucrose (pol/^{0}brix x 100);

   TFS = Total fermentable sugars.
2. Percentage of fresh weight sweet sorghum stems recoverable as ethanol on a mass basis (assumes 35% recovery).

   Ethanol density = 0.789g/L.
3. Expected recovery of ethanol if 46 t stems/ha of sorghum are delivered to the mill.

After preparation of ethanol from sweet sorghum and sugarcane the comparison of per liter cost of production in rupees is given Table 46. As well as comparative economics of sweet sorghum *vs.* sugarcane for common time period is presented in Table 47.

Table 46. Comparison of per liter cost of production (Rs.) of ethanol from sweet sorghum and sugarcane.

Particular	Sweet sorghum	Sugarcane molasses
Manpower	0.50	0.25
Steam	1.00	1.00
Electricity	1.00	1.00
Yeast	0.10	0.10
Management/Admin.	0.10	0.25
Pollution control	Nil	0.25
Raw material	10.41	12.13
Total	13.11	14.98

1. Raw material cost molasses at the Rs. 2000/ton.
2. Raw material cost sweet sorghum at the Rs.500/ton.
3. Alcohol output 48L/ton. from sweet sorghum.

Table 47. Comparative economics of sweet sorghum and sugarcane for common time period.

Parameter	Amount (Rs.)
Sweet sorghum	
(3½ months/per crop, 2 crop cycles is possible) Typical gross returns to farmer: SS stalks @ Rs. 450/t (Two cycles 35 t/ha in I season x 2 cycles) + 35 q/ha of grain @ Rs. 6/kg	52500/ha/year
Minus cultivation cost	31196/ha/year
Net returns	**21304/ha/year**
Sugarcane	
Typical gross returns to farmer: 75 t/ha (Sugarcane @ Rs. 850/t)	63750/ha/year
Minus cultivation cost	45000/ha
Net returns	**18750/ha/year**

Chapter **25**

Utilization of Sweet Sorghum Bagasse

The NARI prepared a gasifier for syrup or jaggery preparation. The gasifier is an open top, throatless (cylindrical) type and requires the biomass residues to be chopped into 1-10 cm long particles, which is carried out using a 2-3 KW electric chaff cutter. The gas produced (mainly comprising of 15-25% carbon monoxide, 10-20% hydrogen and 1-5% methane), is combusted in a special burner designed at NARI. It is basically a venture-type burner with a plate for flame stabilization. It produces a bluish white flame with temperature exceeding 1250 ^{0}C. Gasifiers can also be used for drying of agricultural produce and products and incrematoriums where substantial wood saving can result (Plate No. 15). They used different types of biomass residue in NARI gasifier (Table 48) and found that sweet sorghum stalks and sugarcane leave are excellent for gasification while sweet sorghum bagasse and sugarcane bagasse are very good for gasification.

The SSV 84 is also grown for fodder in some pockets. This has helped to improve milk yield and reduce the cost of concentrates. The sweet sorghum bagasse (the residue after extraction of juice for jaggery or syrup making) is a by-product; it is as good as stover of M 35-1 and can be conveniently used for feeding to animals. The chemical composition of sorghum bagasse of SSV 84 and Sorghum stover of M 35-1 is given in Table 49.

Table 48. Characteristics of biomass residues used in NARI gasifier.

Biomass residue	Yield t/ha/year (oven dry)	Bulk density of chopped material (kg/m^3)	Ash content (%) (w/w dry)	Suitability for gasification
Sweet sorghum stalks	20-30	70-120	4-5	Excellent
Sugarcane leaves	4-6	26-40	7-8	Excellent
Sweet sorghum bagasse	10-14	70-90	4-5	Very good
Safflower residues	3-8	70-80	2-3	Good; increased tar products
Hybrid Napier grass	24-40	42-48	8-12	Fair; problems with bridging of fuel and rapid combustion
Cenchrus grass	5.5-7.0	26-30	11-15	Fair; problems with bridging of fuel and rapid combustion
Wheat husk	4-10	49-53	4.3-11	Good
Sugarcane bagasse	17-20	150-160	1.3-1.5	Very good

Table 49. Chemical constituents of sorghum bagasse (SSV 84) and sorghum stover (M-35-1).

Constituent	Sorghum bagasse	Sorghum stover
Protein (%)	4.5	4.4
Fiber (%)	26.0	28.2
Cellulose (%)	35.0	40.0
Ash (%)	2.8	10.0
Silica (%)	2.0	-
Calcium (%)	0.50	-
Phosphorus (%)	0.10	-

Possible use of Sweet Sorghum Bagasse Fiber for Paper Manufacturing

It has been reported that pulps of sweet sorghum lines can be used for the manufacture of fine quality writing and printing paper as well as corrugated and solid particleboard. The Indian paper industries facing various difficulties mostly for the availability of cost effective fibrous raw materials. There has been drastic shortage of forest-based raw material such as bamboo and pulpwood. The collection and recovery of waste papers in India is largely unorganized and its availability is also limited because of its other various uses. Manufacturing of paper based on straw has remained limited due to the cumbersome and uneconomical task of collection from widely dispersed areas, its storage and its use as cattle feed. Sugarcane bagasse is considered to be a promising alternative to the other raw materials. There was a surplus of bagasse available for paper mills especially from efficient sugar industries. However, recently a massive effort by Government of India to make these factories also produce electricity *via* cogeneration will create shortage of bagasse for paper manufacturing. Under this situation, sweet sorghum may have very good opportunity to supply the raw material to the paper industry and form an alcohol paper complex for getting good returns.

Chapter **26**

Future Lookout for Sweet Sorghum

The review of literature on sweet sorghum indicates that there is scope to commercialization sweet sorghum for syrup/ jaggery/ sugar/ alcohol production in India. For the communalization of sweet sorghum, there is need to strengthen the research and development on following directions.

1. There is a need to develop sweet sorghum cultivars critically for higher green stalks yield and quality characters.
2. Development of cultivars with different maturity periods to spread the supply of raw materials to the sugar industries/ alcohol distilleries.
3. Development of sweet sorghum cultivars with more total sugars and low fibre and soluble starch content.
4. Studies on syrup quality and their storage stability.
5. Studies on staling effect on juice quality of sweet sorghum.
6. Development of suitable agronomic practices to attain good quality yield.
7. Development of sweet sorghum cultivars resistant to diseases and pests as well as to drought or water stress conditions.
8. Problems related to milling, clarification and concentration should be studied thoroughly for increasing efficiency of industrial process.
9. To find out suitable solutions for solving the problems related to alcohol production efficiently from sweet sorghum juice (Fermentation problems).
10. To study the better utilization of by-products for animal

feeding, bio-fertilizers, paper manufactures and bio-coal or other purposes.

Future Prospects:

Exploitation of sweet sorghum crop will be possible with the development of the following points.

1. Use of a mechanical whole stalk harvester.
2. Use of leaf stripping unit, which can be used before or after stalk harvest.
3. Use of high fermentation efficiency yeast strain and improved fermentation methodology and extraction technology.
4. Use of improved cultivars, which are photo insensitive, disease and insect resistant, stress-tolerant, high yielding and new male-sterile lines, which are sweet, thick and tall.

Thus, the total utilization of the sorghum plant in a balanced production of food, feed and selected industrial products will become increasingly important in the developing countries. A total utilization of all components in the sweet sorghum plant for use in the manufacturing and food industries would increase cash flow to the farmer and thereby constitute an incentive for him to increase his production.

Chapter **27**

Ethanol / Alcohol Fuel Production from Agricultural Produce

Alcohol has assumed very important place in the Country's economy. It is vital raw material for a number of chemicals. It has been a source of large amount of revenue by way of Excise Duty levied by State Government on Alcohol liquors. It has a potentiality as fuel in the form of power alcohol for blending with petrol in the ratio of 20:80. Fermentation alcohol has great demand in countries like Japan, United States, Canada and Sri Lanka etc. The systematic alcohol produced by these countries from Naphtha of Petroleum crude is not useful for beverages. Large quantities of alcohol were exported out of country during last few years. The target of alcohol demand as projected in the perspective plan for Chemical Industry, Department of Chemical and Petrochemicals is 3000 million liters per annum by year 2010.

Ethyl alcohol is an important feedstock for the manufacture of chemicals. These chemicals are primarily the basic carbon based products like Acetic acid, Butanol, Butadiene, Acetic anhydride, Vinyl acetate, PVC etc. (Figure 7). Acetic acid and butanol, which are needed in Pharmaceuticals, paints and in many other areas are important industries as they are value added precuts.

Ethylene, ethylene Oxide and Mono-ethylene glycol are also produced from petrochemical route. However, with latest technological development and taking into account the increasing cost of basic petrochemical raw material, it is now possible to produce Ethylene Oxide, Mono-ethylene Glycol etc. starting from ethanol.

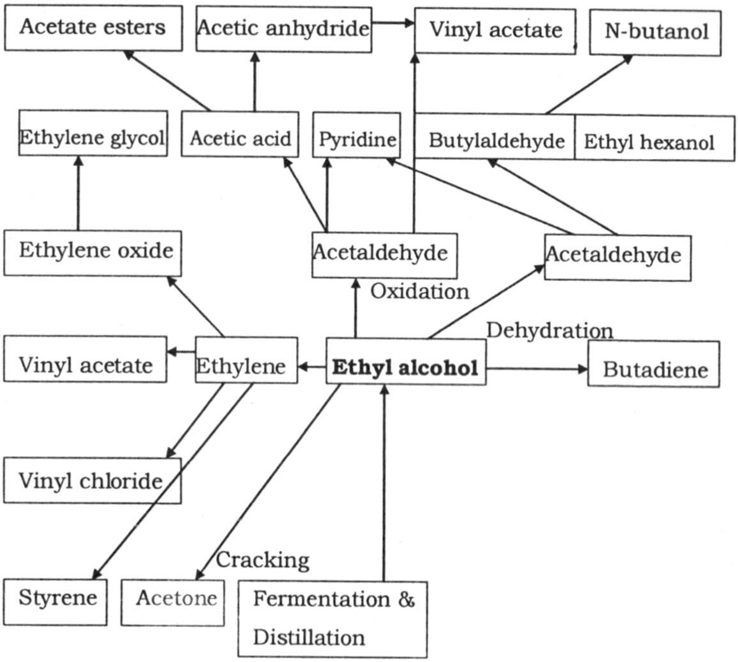

Figure 7. Chemicals from alcohol (Ethyl alcohol).

In special applications, absolute alcohol with 99.8 to 99.9% ethyl alcohol content is required. Therefore,it is necessary to produce absolute alcohol with 99.9% purity and select an appropriate technology. The characteristics of absolute alcohol are given in Table 50.

Table 50. Characteristics of absolute alcohol.

Description	Characteristics of absolute alcohol
Specific gravity at 15.6 ^{0}C	0.7961
Molecular weight	46
Oxygen content	34.70%
Latent vaporization heat	925 KJ/Kg
Lower calorific value	27723 KJ/Kg, 6642 Kcal/Kg
Boiling point	78.30 ^{0}C

Energy per unit and volume	22012 KJ/Kg
Viscosity at 20 ^{0}C	1.192
Vapour pressure at 20 ^{0}C	0.058 Atm
Vapour pressure at 40 ^{0}C	0.177 Atm
Vapour pressure at 60 ^{0}C	0.463 Atm
Octane number (research) RON	106
Octane number (motor) MON	87
Stoichiometric mixture	8.95:1 Air: Fuel

This purified alcohol is called as extra neutral alcohol. It is also useful for cosmetics and perfumes. Absolute alcohol is water free ethyl alcohol and is generally used for purification of pharmaceutical products. The material shall be a clear, clourless and homogenous liquid, free from suspended matter and consisting essentially of ethanol. The process of producing anhydrous alcohol for use as fuel-ethanol is fairly standardized. The technology is available in India and Industrial scale plants are operating, currently for use in the industrial and pharmaceutical applications.

The process of dehydration of rectified spirit entails distillation of the spirit along with an entrainer like hexane. This entrainer takes with itself water present in the spirit, thus rendering the alcohol anhydrous. This water is than separated from hexane in another distillation column, thus recycling hexane.

1. Fuel Ethanol as an Oxygenate

Fuel-ethanol could be used in petrol as an oxygenate. They reduce emission of carbon monoxide in the exhaust gases of vehicles, by taking combustion to completion. It is necessary and advisable to reduce emission of carbon monoxide because it is toxic to human beings. Completion of combustion also reduces emission of particulate carbon matter, which could cause respiratory disorders. Oxygenate has "in-built" oxygen molecule which helps in completing combustion in a better manner. Oxygenates are organic compounds having boiling point in the vicinity of the boiling point of petrol. These compounds mix easily and thoroughly with petrol. Other compounds which are commonly used as oxygenates are tetraethyl lead, MTBE (Methyl-tert-butyl ether), ETBE (ethyl tert-butyl ether), methanol etc. These compounds are also having oxygen molecules in them.

Other function of the compounds added as oxygenates is octane enhancer and anti-knocking agent. These compounds improve the octane number of petrol, thus improving its combustion. Tetraethyl lead is an oxygenate which is rich in oxygen. It has been traditionally used as an oxygenate in petrol. However, it emits poisonous and dangerous fumes containing lead, which are suspected to cause cancer. In order to replace tetraethyl lead, other substances are added to petrol like the aromatics fraction from crude petroleum distillation. This fraction containing BTX-benzene, toluene and xylene is added to improve the octane number. However, emissions from such a mixture are poisonous and require a catalytic converter to prevent such emissions. Thus, vehicles using BTX fractions in petrol have to use a catalytic converter on the exhaust of their vehicle. When MTBE is used as oxygenating agent, there is a danger of it contaminating surface water making it unsuitable for use. In order to get the required content of oxygen, various oxygenates have to be added in different proportion to petrol, based on their content of oxygen.

2. Process of Manufacturing Dehydrated Fuel

Absolute alcohol is an important product required by industry. It is 100% alcohol without any water. Alcohol as manufactured by Indian distilleries is rectified spirit, which is 94.68% alcohol, and rest is water. It is not possible to remove remaining water from rectified spirit by straight distillation as ethyl alcohol forms a constant boiling mixture with water at this concentration and is known as azeotrope. Therefore, special process for removal of water is required for manufacture of absolute alcohol. In order to extract water from alcohol, it is necessary to use some dehydrant or entrainer, which is capable of separating, water from alcohol. Simple dehydrant is unslacked lime, Industrial alcohol is taken in a rector and quick lime is added to that and the mixture is left overnight for complete reaction. It is then distilled in fractionation column to get absolute alcohol. Water is retained by quick lime. This process is used for small-scale production of absolute alcohol by batch process.

3. Dehydration of Alcohol / Ethanol with Entrainer Process

For manufacture of absolute alcohol on large scale, Cyclohexane is used as entrainer. When 94.68% alcohol is mixed with Cyclohexane and distilled a ternary azeotrope is formed. The system consists of two to three columns. First is a dehydration column followed by recovery column. The Rectified spirit is fed into the dehydration column. Cyclohexane is also introduced in this column. Vapour of ethanol, water and cyclohexane close to its azeotropic concentration is collected from the top whereas absolute alcohol is collected from the bottom of the column. Ternary mixture of ethanol, water, and cyclohexane is condensed and sent for decantation where it forms two layers. Top one is cyclohexane rich layer whereas bottom layer is sent to recovery column. Water is collected from the bottom of the recovery column whereas ternary mixture of cyclohexane, water and ethanol comes out of the top, which is condensed and partially sent to dehydration column. Any cyclohexane lost in the system is taken care by adding make-up cyclohexane in the system. One thousand liters of industrial alcohol of 94% V/V contains 940 litres of absolute alcohol and 60 liters of water. For a capacity of 20 KLPD absolute alcohol plant make-up cyclohexane required shall be 50 Kg. The cyclohexane is continuously recovered in the process and recycled as entrainer. About 20 to 30 liters of cyclohexane will be required every day to meet losses of cyclohexane in process. The process is rather simple and well established and given good quality of absolute alcohol.

In another method, glycerine is used as dehyrant. Glycerine is fed counter current wise to the rising alcohol vapours in a column. Glycerine absorbs all the water and leaves from the bottom of the column, which consist of Glycerine, water and some amount of alcohol. Distillate at the top is absolute alcohol glycerine water mixture is sent first to alcohol recovery column and then to vacuum evaporator for recovery of glycerine and removal of water. Glycerine is recycled. This process is also effective giving good quality of absolute alcohol. The dehydration of alcohol / ethanol with entrainer process is presented in Figure 8.

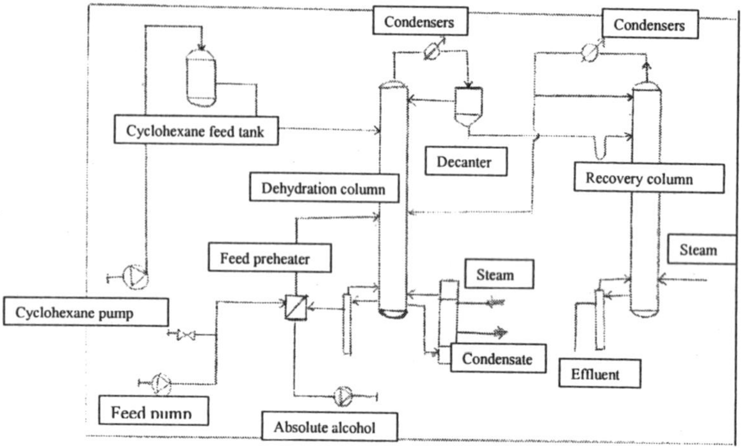

Figure 8. Dehydration of Alcohol / Ethanol with Entrainer Process

4. Dehydration of Alcohol / Ethanol with Molecular Sieve Process

The rectified spirit from the rectifier is superheated with steam in feed super heater. Superheated rectified spirit from feed super heater is passed to one of the pair of molecular sieve beds for several minutes. On a timed basis, the flow of superheated rectified spirit vapour is switched to the alternate bed of the pair. A moderate vacuum is applied by vacuum pump operating after condensation of the regenerated ethyl water mixture. This condensate is transferred from recycle drum to the Rectified Column in the hydrous distillation plant Via Recycle pump. The net make of anhydrous absolute alcohol draw is condensed in product condenser and passed to product storage. The detail flow sheet of alcohol / ethanol dehydration with molecular sieve distillation process is given in Figure 9.

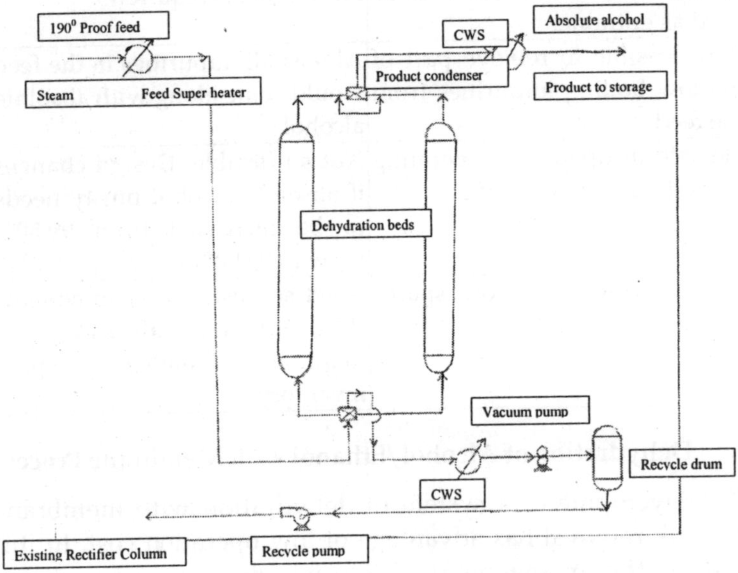

Figure 9. Dehydration of Ethanol / Alcohol with Molecular Sieve Process

Comparative statement of entrainer and molecular sieve process of dehydration of alcohol / ethanol.

Distillation based system	Molecular sieve Dehydration
Lower investment	Higher investment
Higher steam consumption	60% saving in steam consumption Higher power consumption
Degree of automation can vary as desired	Total automatic operation
Long life as distillation is sturdy process	Limited life primarily due to limited life of molecular sieve
Steady and uniform performance	The performance may vary over the period due to deterioration in the molecular sieve.

Solvent cyclohexane needs to be used as entrainer	No solvent is required
It is possible to remove part of the low boiling impurities from the feed	Almost all impurities in the feed would came along with absolute alcohol
Flexible in operation (operating capacity & purity wise)	Not so flexible. Design changes if absolute alcohol purity needs to be increased from 99.50% w/w to 99.90% w/w.
Less moving part, Low spares inventory	Need sophisticated components (PLC, automatic valves etc.) Required higher spare inventory.

5. Dehydration of Alcohol/Ethanol with Membrane Process

The development of system of dehydration with membrane for bulk removal has advantage of low operation cost. In this system, the arrangement of the dehydration and recovery column is the same with the cyclohexane dehydration. The rectified spirit is fed as liquid to a membrane unit. Where water and some alcohol vapour permeates through membrane. The driving force is pressure difference over the membrane. The permeate containing some alcohol flows back to the rectification column. As membrane is expensive, it is advantageous before dehydration to rectify alcohol to strength as high as possible to reduce the membrane cost.

The advantages of this system:

1. Easy operation,
2. Very low energy requirement per liter of absolute alcohol.
3. No toxic or hazardous chemical required.

However, this technology is not yet successfully proved on commercial scale in our country. In this industry, effluent produced is only water. Therefore, problems of pollution hazards are nil. Constant from Environmental Department and Pollution Control Board can be easily obtained as a matter of routine.

6. Bio-ethanol

Ethanol produced from biomass by using microorganisms such as *Saccharomyces cerevisiae* is known as bio-ethanol. The beginning of bio-ethanol use as petrol replacement was began in Brazil and USA during1980s. At present, bio-ethanol is not cost competitive as compared to petrol, but is being used for transport due to government subsidies. Any bio-fuel proposed to be used for transport purpose should have following features:

1. It should be portable in sufficient quantities in the vehicle.
2. It should be burn in the internal combustion engines.
3. It should be roughly equivalent to petrol in energy content.

7. Desirable Characters of Bio-Ethanol Over Petrol

1. Bio-ethanol has much higher latent heat of vaporization (855 MJ/Kg) than petrol (293 KJ/Kg). This energy is obtained from the air in the carburetor. As a result, the fuel mixture entering the cylinder is much cooler and hence denser in case of ethanol than in the case of petrol. Even though ethanol has only 61.8% (27.2 KJ/Kg) energy content of petrol (44 KJ/Kg), the energy produced by combustion of ethanol during each stroke is only slightly lower than that released from petrol.
2. Ethanol has a higher octane number (99) than petrol (80 - 100). As a result, "pre-ignition" does not occur when ethanol is used in engines set for petrol. Pre-ignition denotes ignition before the piston has reached the correct position during compression cycle. Pre-ignition leads to power loss and damage to valves and piston.
3. Ethanol is burnt more completely so that hydrocarbon omission is drastically lower as compared to that in case of petrol.
4. Higher octane rating of ethanol allows the compression ratio of the engines to be increased; results in increased production of power. In view of the above points, ethanol burning engines have only 10% more fuel consumption than

petrol ones although ethanol has only about 62% as much energy content as petrol.

5. Ethanol has a much higher flash point (45 ^{0}C) than petrol (13 ^{0}C). Flashpoint is the temperature at which a substance catches fire. Therefore, ethanol is much less likely to catch fire and explode in cases of fuel leakage (e.g., during accidents).
6. It can be mixed with petrol; this increases the octane rating of petrol. In USA a 20% ethanol and 80% petrol mixture is being marketed as 'gasohol'. However, in such cases the ethanol must be 100% pure otherwise it will not mix in petrol and separation would occur.
7. Alcohol /ethanol is solar energy in liquid form. It can be made in large quantities from renewable sources.
8. The alcohol / ethanol production from agricultural material uses non-protein biomass. All protein can be extracted for human or livestock consumption/usages.
9. Road traffic is the major sources of pollution and produces substantial percentage of Non-hydrocarbons, Co. Today all new are equipped with catalytic converter which are vary efficient in reducing the above pollutants by 90% in ideal conditions.
10. The burning of ethanol / alcohol forms no soot or carbon : carbon build up in an engine is eliminated.

8. Undesirable Characters of Bio-Ethanol as compared to Petrol

1. Engines run on bio ethanol may give starting problems when the air is cool; this is because of the higher latent heat of evaporation of ethanol. This can be overcome by electrically heating the carburetor.
2. Bio-ethanol is highly hydrophilic. As a result, it can absorb moisture from atmosphere etc. presence of water interferes with combustion and also causes corrosion in storage tanks and engines.
3. Bio-ethanol reacts with metals used in the alloys of carburetor etc. e.g. aluminium and magnesium. It can be

remedied by using nickel in the engine alloys.

4. The down stream processing for bio-ethanol recovery is costly, as it requires lots of energy. This is perhaps the chief reason for limiting the use of bio-ethanol as a fuel.
5. Bio-ethanol run engines use about 10% more fuel than petrol, which means a proportionately larger tank.
6. Unlike petroleum product, ethanol / alcohol has no lubricating property, hence additional lubricant like castor oil is required to mix with ethanol / alcohol for proper engine lubrication.
7. When ethanol burns in less quantity of air acetic acid is formed which causes corrosion in engines.
8. Combustion of ethanol / alcohol is likely to be complete due to when range of fuel air ratios, which results in an explosion. With alcohol / ethanol air fuel mixture found in partly empty storage tanks will probably be explosive. So special cellular fuel tank will have to be built for safety.

9. Technical Constraints and Corrective Action Points

A fleet aimed at studying the fuel economy and IIP and IOC (R & D) conducted drivability in case of ethanol-blended fuel and the results of these studies were shown below as blends:

1. Under city driving economy loss was 1% with 10% ethanol blended fuel and 3.90 percent with 20% ethanol blended fuel.
2. Low octane number and high self-ignition temperature makes ethanol / alcohol unsuitable for diesel engines unless the fuel or the engine is modified to match the other.
3. Ethanol / alcohol and diesel are not miscible with each other and thus cannot be used as blend.
4. In gasoline, 100% anhydrous alcohol would be preferred in the long run, which would make the use of blended fuel more uneconomical. A distillery producing ethanol under 1.5 percent water content costs about three times as much as a distillery to produce alcohol of 10% water by volume.
5. A long term continued availability of alcohols will be re-

quired to overcome any meaningful alco-fuel-programme and it has to be fully commercialized to get economical advantage that will require large commercial scale experimentation with alco-fuel, which has not so far been conducted.

6. In the Indian context, no experimentation whatsoever has yet been conducted for finding our operational compatibility of motor engines to ethanol / alcohol /net alcohol as an oxygenator.
7. The production rate of ethanol / alcohol per ton of molasses or other sources is quite low (200 to 240 liters per ton). Efforts should be directed to improve these yields to about 300 to 350 liters. Steps should be taken to produce alcohol from sources other than molasses, such as food grains, sweet sorghum, wood and waste cellulosic materials. Crops with high carbohydrates such as potatoes, corn etc. have theoretical alcohol / ethanol equivalent yield in excess of 2000 liters/hectare.

Sugarcane is indeed one of the most efficient biological converters of solar energy. Its photosynthetic efficiency is 3.8% as compared to earth's average (all vegetation inclusive) of 0.16% only. It has been estimated that 5.6 tons of molasses (approximately 50% sugar) can yield 1 ton of ethanol. Indian being the largest sugarcane country is in a better position with an area of 8.0 million hectares under sugarcane plantation and an average yield of 82.0 tons/hectares at 1990 level besides all agricultural crops and forests can be used to make ethanol / alcohol. Efficient conversion methodologies have to be developed to convert the huge renewable biomass resources base into usable value added products and alternative fuels or conventional fuel blending components for reformulated fuels (Table 51). There is an essential need to formulate the integrated research and development programmes to produce ethanol / alcohol from cellulosic biomass and use the same efficiently for various applications.

Table 51. Comparison of fuel properties of ethanol / alcohols and gasoline

Property	Methanol	Ethanol	Iso-Propanol/ Alcohol	Tert-Butyl Alcohol	Gasoline
Chemical formula	CH_3OH	C_2H_5OH	$(CH_3)_2CHOH$	$(CH_3)_3OH$	C_4 to C_{12}
Molecular weight	32.04	46.07	60.09	74.12	100 - 105
Composition, wt. %					
Carbon	37.5	52.2	60.0	64.8	85.88
Hydrogen	12.6	13.1	13.4	13.6	12.15
Oxygen	49.9	34.7	26.6	21.6	0.0
Sp.gr. 60 ^{0}F /60 ^{0}F	0.796	0.794	0.789	0.791	0.72 - 0.78
Density, lb/gal @ 60 ^{0}F	6.63	6.61	6.57	6.59	6.0 - 6.5
Boiling temp. ^{0}F	149	172	180	181	80.437
Reid vapour pressure, psi	4.6	2.3	1.8	1.8	8.15
Water solubility, @ 60 ^{0}F					
Fuel in water, volume %	100	100	100	100	Negligible
Water in fuel, volume %	100	100	100	100	Negligible
Viscosity, centipoises @ 68 ^{0}F	0.59	1.19	2.38	4.2	0.37 - 0.44
Flash point, closed cup. ^{0}F	52	55	53	52	45
Auto ignition, temp. ^{0}F	867	793	750	892	495
Flammability limits, volume %					
Lower	7.3	4.3	2.0	2.4	1.4
Higher	36.0	19.0	12.0	8.0	7.6
Latent heat of vaporization, Btu/gal @ 60 ^{0}F	3340	2378	2100	1700	-900
Heating value, lower (liquid-fuel-water vapour) Btu/gal @ 60 ^{0}F	56800	76000	87400	94100	109000
Stoichiometric air-fuel weight	6.45	9.00	10.3	11.1	14.7
Ratio moles product/moles $O_2 + N_2$	1.21	1.12	1.10	1.10	1.08

Chapter 28

Bio-ethanol Production

At present, there are three important routes for producing bio-ethanol:

1. From starch or sugar crops,
2. From cellulose following enzymatic hydrolysis and
3. From cellulose following chemical hydrolysis.

The cost of production of bio-ethanol from the three routes is comparable. However, commercial production is based on starch and sugar crops only; there is no full-scale production plant based on cellulose.

Billions of liters of bio-ethanol is produced annually mostly in Brazil (from sugarcane) and USA (from maize). The process based on sugarcane has low productivity and the enormous amount of cellulosic bagasse is either burnt or ploughed back into soil; this biomass is a potential source of bio-ethanol.

Sugarcane, Sugar beet, Sweet sorghum etc. provide a good substrate. Juices from these crops contain sugar, e.g., sugarcane juice has about 12% fermentable sugars, and can be used directly for fermentation. Molasses, sweet sorghum syrup and spent sulphite liquor are the most common substrates. Molasses (C grade) obtained after sugar recovery contains about 35% sucrose, 19% other reducing agents and 14% other organic substances; thus the total fermentable sugars are nearly 50-55%. Molasses is first suitably diluted, then treated with 0.5% by weight concentrated sulphuric acid at 70-95^{0}C (mainly for removal of calcium salts and acidification) and used for fermentation. Cereals like maize, wheat, sorghum etc. contain 60-75% w/w starch, which on hydrolysis produces glucose in the ratio 9:10 (Fig. 10).

Starch
(e.g. powdered seed meal, tubers etc.)

↓ ← Enzyme hydrolysis (50-60 °C)

Glucose

↓ ← Yeast Glycolysis (30-35 °C)

Pyruvate

↓ ← Yeast (30-35 °C) (Pyruvate decarboxylase

Acetaldehyde

↓ ← Yeast (30-35 °C) Alcohol dehydrogenase

Ethanol
(Recovered by distillation)

Figure 10. Flow chart for the production of ethanol from a starch substrate.

Generally, starch is a mixture of amylose (20-30%; water soluble linear polymer) and amylopectin (70-80%; water insoluble branched polymer). A cooking process aided by enzymes usually achieves conversion of starch into glucose. Seeds are powdered and the resulting meal is mixed with water in the ratio of 1:2.5 - 3. α-amylase is then added and the temperature is raised by steam injection to 105-110 °c and held for about 20 min. During this period, starch grains become dissolved to form a viscous suspension (gelatinization) due to the cooking effect of heat. At the same time α-amylase digests partially some of the starch molecules; this reduces the viscosity of the suspension; this is called liquefaction. The suspension contains solubilized starch molecules, and dextrins (short chains of glucose units). The suspension called mash is cooled to 85-90 °c and more α-amylase is added to complete the process of liquification (for 90 min).

The total dose of α-amylase is about 1.5kg/ton starch. The mash is cooled to 55-60 ^{0}C and 1.5-3.5 liter glucoamylase/ton of starch is added to produce glucose from dextrins and maltose. The reaction duration is usually 2 days, and it achieves the dextrose equivalent of 99. Saccharification by glucoamylase continues during the fermentation.

Proper fermentation of 100g glucose by selected strains of *Saccharomyces cerevisiae* and *Saccharomyces carlsbergensis* yields 45-49g bio-ethanol, the theoretical limit being 51.1g.

($C_6H_{12}O_6 \longrightarrow 2C_2H_5OH + 2CO_2$). For the fermentation process batch or continuous processes are used. Batch process is the most commonly used. Ammonium sulphate or Urea (N source) and a salt of phosphoric acid (P source) are added to the substrate. Most commonly the temperature during fermentation is 32-38^{0}C and pH is between 4.5-5.0.

As the ethanol level rises, both ethanol production and yeast cell growth are progressively inhibited, and eventually cells begin to die.

In general, yeast growth stops at 6-9% w/v ethanol, but ethanol production, at least in some strains, may continue upto 15% ethanol or higher. Ethanol inhibition can be relieved by continuous removal of ethanol. Ethanol recovery is based on distillation.

a. The broth is distilled in a beer column to yield 85% v/v ethanol.

b. The next step of rectification gives 96.5% ethanol, which is then,

c. Dehydrated to 99.4% using benzene or cyclohexane if the ethanol is to be used as a fuel blend.

The wastewater effluent can be used directly on land, used for anaerobic digestion for methane generation, evaporated to dry syrup for animal feed or fuel or used as substrate for fungal biomass production.

Chapter 29

Future Lookout for Bio-ethanol Production

1. To increase the substrate utilization ability of excellent ethanol produces like yeast and *Zymomonas mobilis.*
2. To enhance alcohol tolerance of yeast.
3. To increase the bio-ethanol production ability of bacteria capable of utilizing cellulose, hemicellulose, pentoses etc.
4. Practical exploitation of the superior capabilities of *zymomonas mobilis.*
5. Efficient and cost effective process of continuous recovery of ethanol needs to be developed.
6. More efficient and cheaper methods of ethanol recovery need to be developed to reduce the cost of recovery. Some possible approaches may be reverse osmosis, selective adsorption using solid adsorbents and use of supercritical CO_2 to selectively extract ethanol.

Chapter **30**

Biotechnology in Bio-ethanol Production

The potential areas where biotechnology can make a notable contribution are:

a. Development of more efficient organisms for alcoholic fermentation,

b. Devising a more efficient use of these microorganisms e.g. in cell immobilized systems,

c. To develop organisms capable of utilizing cellulose and hemicellulose present in corn and other sugar crops and

d. Practical exploitation of the advantageous features of *zymomonas mobilis*.

Recovery of Bio-ethanol from Biomass

Recovery of ethanol from the fermentation broth is by distillation, which exploits the difference in boiling points of ethanol (87^{0}C) and water (100^{0}C). As a result, when a water ethanol mixture having 95% ethanol is heated, the vapour has a greater concentration of ethanol than the liquid phase. Thus a dilute ethanol : water mixture can be repeatedly distilled to obtain more concentrated (up to 95%) ethanol solution. The principle of sequential distillation is used in a cylindrical distillation column, which is divided into a series of chambers by perforated plates. The ethanol water mixture is boiled using steam and the vapour rises into the column. Essentially, each chamber of the column functions as a distillation unit so that the proportion of alcohol goes on sequentially increasing as it rises to the upper chambers of the column. A properly designed column would yield 95% ethanol from its top most chamber.

Such a column can be run either in a batch or in continuous mode.

When 95% ethanol is heated, the proportion of water present in the vapour phase is the same as that in the liquid phase. Therefore, simple distillation is no more helpful in further purification/concentration of ethanol. A small amount of benzene is added to 95% ethanol, which is then distilled. The distillate is nearly 100% ethanol, and the benzene can also be recovered. Thus recovery of pure ethanol requires considerable amounts of energy, which raises its production cost; this has been the chief reason for the limited use of this valuable biofuel/ bio-ethanol for transport purposes.

Chapter 31

Industries and Firms

Ethanol Producing Industries from Sweet Sorghum on Pilot Scale

1. M/s, Sri. Renuka Sugar Factory
   Munoli, Belgaum District,
   Karnataka.
2. M/s, Ethanol and Chemical Industries
   At. Pangarkhel
   P.O. Budh
   Tal. Khatav
   Dist. Satara.
3. M/s, Nimbkar Agricultural Research Institute (NARI)
   P.O.Box. 44, Phaltan, Tal. Phaltan, Dist Satara.
4. M/s, Praj Industies Pune
   Dhankawadi Pune
   http:/www.prdomain.com.
5. M/s, Sancorn Agro processing Ltd. Kolhapur
   612 B, Anandi Chembers
   Shahupuri, 2nd Lane
   Kolhapur- 416001.
   http://www.sancorn.com.
6. M/s, Sagar Sugar and
   allied Chemicals Chittoor.

Sorghum Fodder Bricks Manufacturing Firms

1. M/s. Mangalweda Kadaba-Dryfodder,
   Research and Development Corporation,
   Mangalweda, Solaphur.
2. M/s. Miracle Fodder and Feeds Private Ltd.,
   Survey No. 674, N.H.-7, Shamshabad,
   Dist: Rangareddy, Hyderabad - 500 012.

Chapter 32

References

Anderson, I.C. 2005. Ethanol from sweet sorghum. IOWA Energy Center. Online publication. 1.

Anderson, I.C. 2005. Ethanol from sweet sorghum. http://www.energy. iastate.edu/renewable/biomass/csanerobicz.html

Anonymous, 2002. Annual Research Report, NRCS, Hyderabad.

Anonymous, 2002. Annual Research Report. Sorghum Project, MPKV, Rahuri.

Annual report on sorghum, 2002-2003, AICSIP, NRCS, Hyderabad. Pp.87.

Annual report on sorghum, 2002 - 2003. Research Review Committee Meeting, MPKV, Rahuri. pp. 126.

Annual report on sorghum, 2007. Research Review Committee Meeting, MPKV, Rahuri. pp. 150.

Bapat, D.R. and S.D. Chaudhari. 1976. Screening sweet sorghum varieties for sugar production. Newsletter. 19 : 47.

Bapat, D.R., C.B. Salunke, H.D. Jadhav and S.P. Patil. 1988. Breeding for improvement of sweet sorghum. Paper presented at XVIIII Annual Sorghum Workshop at C.S. Azad University of Agriculture and Technology, Kanpur during 2-4 May, 1988.

Chaintawongvanich, S., J.P. Clark and M.E. Wright. 1977. Rapid evaporator for sorghum syrup. AICUE Symposium Series. 73 : 191-197.

Chatterji, A.C., S.N. Joshi, K.B. Nadgauda, S.R. Kalaswad, B.R. Math, V.P. Taware, 1975. White sugar production from sweet sorghum. S.T.A.I., Golden Jubilee Workshop, Kanpur. 73-80.

Clark, J.W., F.R. Miller and R. Creelman. 1980. Utilization of sorghum biomass for energy. Sorghum Newsletter. 23 : 44-45.

Coleman, O.H. 1970. Syrup and sugar from sweet sorghum. Sorghum production and utilization. The AVT publishing Company, Inc., Westport, Connecticut, USA. pp. 416-440.

Coleman, O.H. and D.M. Breadhead, 1968. Brandes a new variety of sweet sorghum for syrup production. Mississippi State University Agril. Extension Station. Information sheet 1013. State College Mississippi, U.S.A. Mississippi State University.

Coleman, O.H. and I.E. Stokes, 1958. Inheritance of weak stalk in Sargo. Agro. J. 50 : 119-120.

Coleman, O.H. and I.E. Stokes, 1964. Storage studies of Sargo. U.S. Department Agro. Tech. Bull. 1307.

Dogget, H. 1970. Syrup, Sargo and Sudangrass. Sorghum.

Doggett, H. 1988. Sorghum. Harlow, UK, Longman Scientific and Technical. Pp. 512.

Energy Authority of NSW. 1986. Sugarcane to fuel Ethanol: Feasibility study. Sydney, Australia. Energy Authority of NSW. ISBN 0-7305-2799-9:1-159.

Ferrais, R. 1988. Sweet Sorghum: Agronomic Evaluation and potential Application for Australian Agro-Industry. 5, Research Report, CSIRO (commonwealth scientific and industrial research organization), Brisbane. Australia. Pp. 1-49.

Food and Agriculture Organization (FAO). 2000. FAOSTAT Database, http://www.fao.org

Freeman, K. C., Broadhead, D. M., Zummo, N. and Westbrook, F. E. 1986. Sweet sorghum culture and syrup production. U.S.Dept. of Argil. Handbook No. 611. pp.55.

Freeman, K.C., Broadhead, D.M., Zummo, N. and Westbrook, F.E. 1986. Sweet sorghum culture and syrup production. U.S. Dept. of Agr. Handbook No. 611. pp. 55.

Hallgren, L., Rexen, F., Peterson, P. B., and Munck, L. 1992. Industrial Utilization of whole crop sorghum for food and industry. Pages 121-130. In: Utilization of sorghum and millets: Proceeding of the international workshop on policy, practice and potential relating to uses of sorghum and millets. 8-12.

Hills, F.J., S.S. Johnson, S. Geng, A. Abshashi and G.R. Peterson. 1981. Comparison of high energy crops for alcohol production, California Agriculture. 35 : 14-16.

Jeremy Woods, 2001. The potential for energy production using sweet sorghum in Southern Africa. Energy for Sustainable Development. 1 : 31-38.

Jorapur, R. M. and Rajvanshi, A. K. 1995. Development of a sugarcane leaf gasifier for electricity generation. Biomass and Bioenergy. 8 (2): 91-98.

Li, D. 1997. Developing sweet sorghum to accept the challenge of problems on food, energy and environment in 21st century. Li, D., (ed), Proc. First International sweet sorghum conference, Chinese Academy of sciences, Beijing. China. Sep. 25, pp. 19-34.

Mask, P. L. and Morris, W. C. 1991. Sweet sorghum culture and syrup production. Circular ANR-625. Alabama cooperative Extension service, Auburn University.

Morris, J.B. 2005. Production of sweet sorghum for syrup in Kentucky. Online publication. AGR-122. 1-5.

NARI, 1988. Ethanol production from sweet sorghum. A final project report submitted to the Department of Non-conventional Energy Sources, New Delhi, by the Nimbkar Agricultural Research Institute, Phaltan. pp. 59.

NARI, 1995. Development of sweet sorghum (*Sorghum bicolor* (L.) Moench) lines giving high stalk yield and good quality juice for production of industrial ethyl alcohol. 1995. A final project report submitted to the Ministry of Non-conventional Energy Sources, New Delhi by the Nimbkar Agricultural Research Institute, Phaltan. pp. 83.

Phule Amruta, 2003. Mahatma Phule Krishi Vidyapeeth, Rahuri. University Extension Publication. No. 408.

Rajvanshi, A. K. and Jorapur, R. M. 1993. Development of leafy biomass gasification systems. Publication No. NARI-GS-1, published by Nimbkar Agricultural Research Institute (NARI), Phaltan, MS, India.

Rajvanshi, A.K., De T.K., Jorapur, R.M. and Nimbkar, N. 1993. Jaggery and syrup from sweet sorghum. Publication No. NARI-GUR, Published by Nimbkar Agricultural Research Institute (NARI) Phaltan, M.S., India.

Rajvanshi, A.K. and Nimbkar, N. 2005. Sweet Sorghum Research and Development at the Nimbkar Agricultural Research Institute (NARI). 29-34.

Rajvanshi, A.K. and W. Nimbkar, 2005. Sweet sorghum R & D at the Nimbkar Agricultural Research Institute (NARI). Source Internate.

Rajvanshi, A.K., R.M. Jorapur and N. Nimbkar. 1994. Sweet sorghum R & D for food and fuel. pages 29-34 in to explore the possibilities of cultivation and processing of supplementary sugar crops : proceedings of VII Joint Convention of the Deccan Sugar Technologists Association (DSTA) and the Sugar Technologists Association of India, DSTA, Pune - 411 005, India.

Ratnavathi, C.V. 2002. Annual Report : Developing Sorghum as an efficient biomass and bio-energy crop and providing value addition to the rain damaged *kharif* grain for creating industrial demand. 1-27.

Reeves, S.A. and B.A. Smith, 1979. Sweet sorghum biomass a renewable energy source Part-2. Technical report 79-1 College Station, Texas, USA. Texas Agricultural Experiment Station, Texas A & M University.

Roman, G. V. 1995. Bioethanol production from sweet sorghum: Interchange of research experience between EC and two east European Countries (Romania and Hungary) Romania Report. EU DGXII PECO Project Reports, Brussels, EU.

Rosillo-Calle, F. 2000. Pres, com., The value of anhydrous ethanol in Brazil, London. Dec.

Rothman, H., Green shields, R. and Rosillo-Calle, F. 1993. Energy from Alcohol: the Brazilian Experience, University of Kentucky Press, Lexington, Kentucky. USA. Pp. 1-188.

Schaffert, R. E. and Gourley, L. M. 1982. Sorghum as an energy source. Pages 605-623. In: sorghum in the eighties: Proceedings of the International Symposium on sorghum, 2-7 Nov. 1981, ICRISAT Center, India. Patancheru, A.P. 502324, India. International Crops Research Institute for the Semi-Arid Tropics.

Schaffert, R. and L.M. Gourly, 1982. Energy sources, In sorghum in the Eighties, Proc. of an Intl. Symp. NSV. 1-7, 1981. ICRISAT Centre, Patancheru- 502 324, India.

Schaffert, R.E. and R.A. Borgonovi. 1980 b. Prospectivas de sargo sacrino para producaode ethanol. In press in Simposio Soore alcohol. Academia de ciencies do Estadode Sao Paulo, Brazil.

Schaffert, R.E. and R.A. Borgonovi. 1980a. Sorgouma occao para a producaode alimentos a energia. A cranja. 36 : 60-64. Edibora Centa Urus Ltda, Porto Alegre, Brazil.

Undersander, D.J., Lueschen, W.E., Smith, L.H., Kaminski, A.R., Doll, J.D., Kelling, K.A. and Oplinger, E.S. 1990. Sorghum for syrup. Alternative Field Crops Manual. p. 1-7.

Undersander, D.J., W.E. Lueschen, L.H. Smith, A.R. Kaminski, J.D. Doll, K.A. Kelling and E.S. Oplinger. 2005. Sorghum for syrup. Alternative Field Crops Manual.

Vance, J.E. 1980. Sorghum : Food + Feed + Fuel, Seeds and Farms. 6: 41-42.

Venkatachari, J. 2004. Praj Industries cultivates sweet sorghum to make blended fuels. WWW.domain-com/ companies/ companies_P/Praj/ 20041030-cultivates.html

Ventre, E.K. and S. Byall, 1939. Distribution and variation with maturity of dissolved solids, sucrose and titratable acidity in the Sorgo stalk. J. Agric. Res. 55 : 553-556.

Ventre, E.K., S. Byall, Jr. C.F. Waltan, 1939. Jellying and crystallization of syrup made from different parts of the sargo stalks at different stages of maturity. J. Agric. Res. 59 : 139-150.

Wankhede, D.B., Syed Ismail, Shinde, G.G. and Ambekar, S.S. 2005. Sorghum : An efficient and a potential crop for industrial utilization. Technical bulletin on sorghum. MAU, Parbhani. pp. 12.

Wenman, C. (Technical Director, Triangle Ltd. Zimbabwe), 1999. Pres.comm.., Potential for crystalline sugar production from sweet sorghum.

Wilhelms, L. R. and McCarty, I. E. 1985. An evaluation of sorghum syrup processing operations in Tennessee. Research Report 85-01. Agricultural Experiment Station, University of Tennessee.

Woods, J., Hall, D. O., Muzondo, M. I., Gosse, G., and Soontronchainackseng, P. 1995. Bioethanol production from sweet sorghum: Interchange of research and experience between EU and Developing countries (Zimbabwe and Thailand), Brussels, EU, pp. 1-74.

Woods, J. 2000. Integrating sweet sorghum and sugarcane for Bioethanol: Modeling the potential for electricity and ethanol production in SE Zimbabwe, PhD thesis, King's College London.

Woods, J. 2001. The potential for energy production using sweet sorghum in Southern Africa. Energy for sustainable Development. 5(1): 31-38.

APPENDEX - I

Conversion factors (weights and measures)

Length	1 m	10^{10} A	39.37 in	3.2808 ft
Area	1 m^2	10^4 cm^2	1550.0031 in^2	10.7639 ft^2
Volume	1 m^3	1000 L	10^6 cm^3	264.17 gal (US)
Mass	1 kg	2.2046 lbm	35.274 oz	1.1023 x 10^{-3} ton (short)
Density	1 g/cm^3	1000 kg/m^3	62.428 lbm/ft^3	8.3454 lbm/ gal (US)
Velocity	1 m/s	3.6 km/h	2.2369 min/h	196.85 ft/min
Force	1 kg-m/s^2 (N)	0.2248 lbf	10^5 g.cm/s^2 (dyn)	7.2330 lbm.ft/ s^2 (pdf)
Pressure	1 atm	101.325 kg/ms^2 (pa)	14.696 lbf/in^2 (psi)	760 mm Hg
Energy	1 kg.m^2/s^2 (J)	9.4782 x 10^{-4} Btu	0.2388 cal	0.7376 ft.lbf
Power	1 J/S (W)	1 kg/m^2/s^3	3.4121 Btu/h	1.341 x 10^{-3} hp
Thermal conductivity	1 W/m.k	0.5778 Btu/h.A.^{0}F	2.388 x 10^{-3} cal/s.cm.^{0}C	6.9335 Btu.in/h.ft^2.^{0}F
Specific heat	1 KJ/kg.K	0.2388 cal/g.^{0}C	0.2388 Btf/lbm.^{0}F	185.86 ft.lbf/lbm.^{0}F
Viscosity (dynamic)	1 pa.S	1 kg/m.s	1000 cp	0.6720 lbm/ft.S
Temperature	^{0}C=5/9 (F-32)	^{0}F = 9/5 (^{0}C) + 32	K = ^{0}C + 273.15	^{0}R = ^{0}F + 45967

APPENDEX - II

Agricultural Related Websites:

http://www.nrcsorghum.res.in
http://www.aau.ac.in
http://www.mpkv.nic.in
http://www.hau.nic.in
http://www.csauk.org
http://www.pdkv.mah.nic.in
http://www.gud.guj.nic.in
http://www.mpkv.mah.nic.in
http://www.hpkv.nic.in
http://www.icar.org.in/sherk/welcome.htm
http://www.gny.ifas.uf.edu
http://www.agri.surf.com
http://www.agfind.com
http://www.morrisville.edu/njc
http://www.agr.ca/agnweb/agmweb.htm
http://www.agri.infolink.com
http://virtualquincy.com/quincy/ag/agricultural.
http://www.icar.org.in
http://www.iaripusa.org
http://www.cishlko.org.
http://www.cri.ogr
http://www.cpri.hp.nic.in
http://www.cpri.nic.in
http://www.kar.nic.in
http://www.kar.nic.in/cashew
http://www.nic.map.org
http://www.hocmushroom.org
http://www.cazri.nic.in
http://www.ndri.nic.in

http://www.nbsslup.org
http://www.nbpgr.delhi.nic.in
http://www.nic.ipm.org
http://www.dor_icar.org
http://www.icar_org.in/dwr/dwrmain
http://www.cswri.raj.nic.in
http://www.cirg.nic.in
http://www.cicr.india.org
http://www.rice.india.tripool.com
http://www.upgov.up.nic.in
http://www.amazon.com
http://www.servtech.c. (Digital Library)
http://www.leweb.gov/. (The Library of Congress)
http://www.Thorplus.Lib.purdue.edu/vlibrzy(The virtual Library)
http://www.nal.usda.gov./NAL (National Agricultural Library)
http://www.Fao.hg/Library/Librs.htm. (FAO-Library)
http://www.Cobweb Common Wealth Agricultural Bureau
http://www.ioe.org - Journal of Extension
http://www.progressive farmers.com – Progressive farmers
http://www.agecon.ugd.edu/jab-Journal of Agribusiness
http://www.agecom.uga.edu/jade/-Journalof Agricultureand applied economic
http://www.agrivareers.com
http://www.agrilaunch.com
http://www.agriculture online.com
http://www.cgiar.org
http://www.kisan.net
http://www.myrada.org
http://www.nads.edu
http://www.fao.org

http://www.wellcent
http://www.apeda.com
http://www.falbazar.com
http://www.herbal.com
http://www.ecobot.coms
http://www.agri.gov.ab'ca/crops/badimarketing.com
http://www.hortibizindia.nic.in
http://www.mofpi.nic.in
http://www.farmersshaw.net/chei/coneers
http://www.agnet.co.uk.agsan.htm
http://www.infolink.com
http://www.aginfonet/sk.ca
http://www.msamb.com
http://www.aqriwatch.com
http://www.agmarket.nic.in

APPENDIX – III

About Sorghum Improvement Project, Mahatma Phule Krishi Vidyapeeth, Rahuri.

The Sorghum ImprovementProject is started at Mahatma Phule Krishi Vidyapeet, Rahuri in 1973 with following major objectives:

1. To develop new genotypes and hybrids in *rabi* sorghum.
2. To develop advanced agricultural technology for cultivation of *rabi* sorghum for higher yield and quality.
3. To develop sorghum genotypes for alternative uses such as sweet sorghum for ethanol production and other alternative uses.

Scientists working at Sorghum Improvement Project have developed various new *rabi* and alternative uses sorghum genotypes as given below:

New Genotype	Year of release	Grain yield (q/ha)	Fodder yield (q/ha)	Remarks/ Recommended
A. New genotypes for grain and fodder purpose				
Swati	1984	24-28	50-60	Dryland/Irrigated
Selection - 3	1994	5-6	15-18	Shallow soil
Phule Yashoda	1998	20-25	70-75	Irrigated area
Phule Maulee	1999	8-9 15-20	20-25 50-60	Shallow soil Medium soil
Phule Chitra	2006	20-25	55-60	Medium soil
Vasudha	2007	30-35	70-75	Deep & irrigated soil
CSV - 22	2007	24-28	65-70	Deep soil, irrigated
Anuradha	2008	8-10	30-35	Shallow, dryland condition
RSV 1006	2009	40-50	90-100	Irrigated, Deep soil
B. New genotypes for alternative uses				
SSV 84	1991	10-12	300-350 (green)	Sweet sorghum for ethanol
CSV 19 SS	2004	10-15	350-400 (green)	Sweet sorghum for ethanol
Phule Amruta	2003	-	450-500 (green)	Green fodder, Irrigated condition
Phule Uttara	2005	20-25	50-60	For Hurda
RPOSV 3	-	12-15	30-35	For pops
RPASV 3	-	8-10	30-35	For papad

APPENDEX – IV

Websites for Nutritional Information

http://www.users.rcn.com/jkimball.ma.ultrnet/BiologyPages/N/Nutrition.html

http://www. Ars.usda.gov/ba/bhnrc.

http://www.asinut.net

http://www.globnut.net

http://www.nutrition.otago.ac.nc

http://www.mrc-hnr.cam.aguk/

http://www.cumc.cdumbia.edu/deptlihn

http://www.faoorg/docrep/W8079E/W8079E00.htm

http://www.fshn.uiuc.edu

http://www.blackwellpublishing.com/journals/jhn

http://www.COREgroup.org

http://www.DevelopmentGateway.org

http://www.suhrs.dk

http://www.MonsterIndia.com

http://www.hnrc.tufts.edu

http://www.ars.usda.gov/research/programs/programs.htm?NP_CODE=107

http://www.grad.ubc.ca/apply/programs

http://www.gla.ac.uk/departments/humannutrition/index.html

http://www.uchsc.edu/nutrition

http://www3.open.ac.uk/courses/bin/p12.

http://www.stfx.ca/academic/human-nutrition

http://www.itana2002.org

APPENDEX – V

PROJECT REPORT (STANDARD FORMAT)

OF

PROCESSING INDUSTRIES

1. Primary information

a. Name of Unit : ....
b. Units' address ....
c. Constitution :..
d. Name of proprietor :..
e. Birth date, Age :..
f. Qualification :..
g. Nature of business :..
h. Registration No. of business :....................................
i. Loan for :..
j. Loan amount :..

2. Introduction

In this section the information on following points should be included:

a. Experience of the proprietor in this field
b. Facilities and arrangement for product sale
c. Information regarding machinery requirement and their availability as well as importance
d. Availability of the skilled and unskilled labour
e. Information regarding self-sale counter
f. Importance of this business to the society

3. Scope of business

Give the location of the processing unit and availability of raw material in that area. Provide the information regarding marketing of the processed product.

4. Manufacturing process

Give the details of manufacturing process/technology employed.

5. Land

Land for proprietor owns unit or it is in partnership give details. If in partnership then make agreement and submit with this proposal.

6. Building

Give the details about building construction (shed, RCC or load bearing etc.).

7. Water supply

Water supply from bore well or Municipal Corporation etc.

8. Electric power supply

MSEB connection deposit, main switchboard, wiring, fitting, tubes, fans, labour charges etc.

9. Printing

For labeling, advertisement, specification on the product etc.

10. Furniture

Expenditure occurred on tables, chairs, cupboards, stools etc.

11. QUANTITATIVE DETAILS

(Rs. In Lacs)

Sr. No.	Description	1St year	2nd year	3rd year	4th year	5th year
1.	Installed capacity 300 days & 20 hrs					
2.	Utilization of capacity (%)					
3.	Raw material consumption at installed capacity (Rs./Ton)					

4.	Sale at installed capacity (Rate Rs. / Ton)					
5.	Raw material consumption at utilized capacity					
6.	Sales at utilized capacity					

12. CALCULATION OF WORKING CAPITAL

Sr. No.	Particular	Holding period (days)	Amount
1.	Raw material		
2.	Consumable packing		
3.	Finished goods		
4.	Other Exp.		
5.	Debtors		
Gross current assets:			
6.	Less credits		
	Net working capital		
	Own margin	35 or 25%	
	Bank Finance	56 or 75%	

13. COST OF PROJECT AND SOURCES OF FINANCE

Sr. No.	Item	Cost already incurred	Cost to be incurred	Total
A. Cost of project:				
1.	Land			
2.	Site development			
3.	Civil & Technical work			
4.	Plant & Machinery			
5.	Total working capital			
6.	Preliminary & preoperative Exp. (Bank fee, mortgage fee, Prof. Fee etc.)			
	Total			

B. Source of finance:				
1.	Equity capital			
2.	Subsidy/Quasi equity			
3.	Others			
4.	Unsecured Loan			
5.	Term Loan			
	Total			

14. PROJECT PROFIT AND LOSS ACCOUNT

(Rs. In Lacs)

Sr. No.	Particular	1st year	2nd year	3rd year	4th year	5th year
1.	Gross sales					
2.	Job work charges					
A.	Net sales					
B. Cost of production						
1.	Raw material (as per qty. details)					
2.	Power & Fuel					
3.	Direct labour & Wages					
4.	Consumables stores					
5.	Repair & Maintenance					
6.	Other Mfg. Exp.					
7.	Depreciation					
8.	Add opening stock in process & finished goods					
9.	Deduct closing stock in process & finished goods					
C	Cost of sales					
D	Gross profit (A- C)					
E	Interest					
F	General & administrative over					
G	Profit before Tax					

H	Provision for Tax					
I	Net Profit					
J	Depression added back					
K	Net cash accruals					
L	Repayment obligations					
M	Debt service ratio (Average DSCR 3.87)					

15. PROJECTED BALANCE SHEET

Sr. No.	Particular	1St year	2nd year	3rd year	4th year	5th year
A. Assets						
1.	Fixed assets					
2.	Investment/L. C. Deposits					
3.	Current assets					
	a. Raw material, packing Exp.					
	b. Finished goods					
	c. Debtors					
	d. Other current assets					
	Total					
B. Liabilities						
1.	Equity capital					
2.	Profit & loss A/C					
3.	Unsecured Loans					
4.	Quasi equity					
5.	Subsidy (Fr. MFPI)					
6.	Secured Loans					
7.	Working limits Fr. Bank					
1.	Term Loan					
2.	Creditors					
3.	Other current Liab. & Provision					
	Total					

16. PROJECTED CASH FLOW AND FUND FLOW STATEMENT

Sr. No.	Particular	1St year	2nd year	3rd year	4th year	5th year
A. Sources						
1.	Net cash accruals					
2.	Subsidy					
	Total					
B. Application						
1.	Repayment of term Loan					
2.	Repayment of unsecured & quasi equity					
3.	Increase in investment					
4.	Increase in W. C.					
	Total					

17. PROJECTED BREAKEVEN

(Rs. In Lacs)

1.	Fixed cost (Average)		
2.	Average selling price		
3.	Average variable cot (per ton)		
	a. Materials		
	b. Labours		
	c. Other overheads		
4.	Contribution/Unit		
5.	Breakeven qty		
6.	Breakeven value		
7.	Percentage on installed capacity		

18. REPAYMENT SCHEDULE OF TERM LOAN & INTEREST ON TERM LOAN

Period	Principle amount	Interest	Total
Interest in ....%, period in quarters			
1st Year			
I Quarter			
II Quarter			
III Quarter			
IV Quarter			
Total			
2nd Year			
I Quarter			
II Quarter			
III Quarter			
IV Quarter			
Total			
3rd Year			
I Quarter			
II Quarter			
III Quarter			
IV Quarter			
Total			
4th Year			
I Quarter			
II Quarter			
III Quarter			
IV Quarter			
Total			
5th Year			
I Quarter			
II Quarter			
III Quarter			
IV Quarter			
Total			
Grand Total			

19. CALCULATION OF DEPRECIATION AS PER I. T. RULES & W.D.V. METHOD

Period	Building	Plant and Machinery	Total
Initial			
1st Year			
Total			
2nd Year			
Total			
3rd Year			
Total			
4th Year			
Total			
5th Year			
Total			

20. LIST OF PLANT AND MACHINERY

Sr. No.	Name of Plant or Machinery	Price (Rs.)	Quantity (Nos.)	Total amount (Rs.)
1.				
2.				
3.				
4.				
5.				
6.				
7.				
8.				
9.				
10.				
	Total			

21. TOTAL INTEREST PAYABLE YEARWISE

Period	Term	Working	Other	Total
@ p. a.	11%	15%	18%	
1st year				
2nd year				
3rd Year				
4th Year				
5th Year				
Total				

22. SALARY AND WAGES DETAILS

Sr. No.	Particular	Nos.	Salary p.m.	Total
1.	Operator			
2.	Skilled workers			
3.	Unskilled workers			
	Total			

Proprietor/Partner
(Name & Address of the Unit)

Plate No.1 Soil moisture conservation technology

Plate No. 2. Hand hoeing interculturing operation

Plate No. 3. Midge fly

Plate No. 3. Stem borer

Plate No. 3. Stem borer

Plate No. 4. Checking ^{0}Brix reading

Plate No. 5. Mature sweet sorghum canes and standing crop

Hydro Pack 40 HP

Sorghum Fodder Bricks

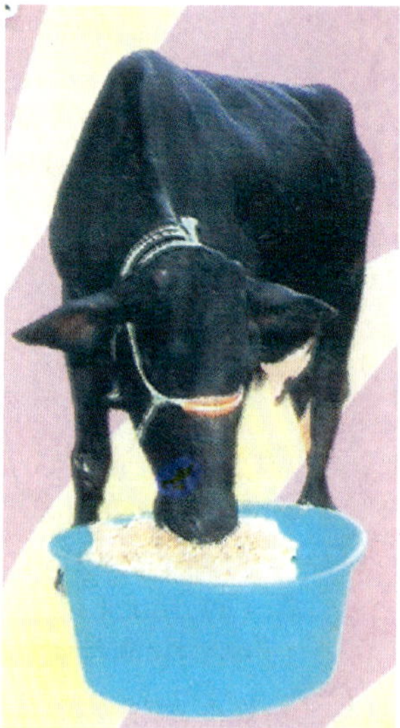

Sorghum Bricks feeding

Sorghum Bricks storage

Plate No. 6. Sorghum Fodder Bricks Preparation, Feeding and Storage

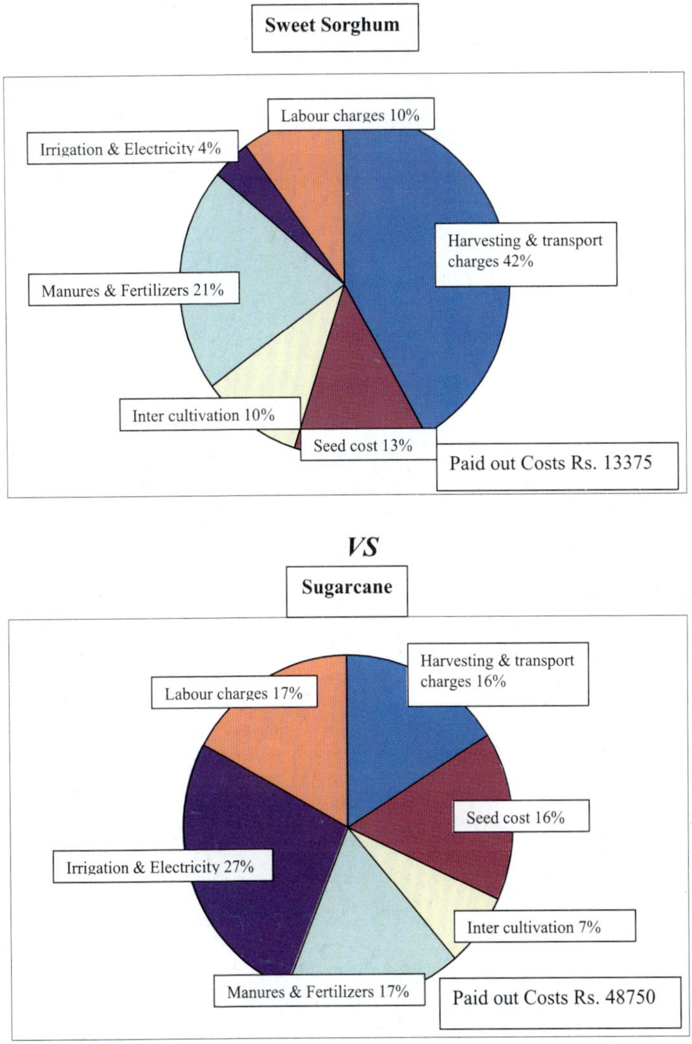

Plate No. 6. Item-wise paid-out costs of Sweet Sorghum *VS* Sugarcane cultivation in Maharashtra.

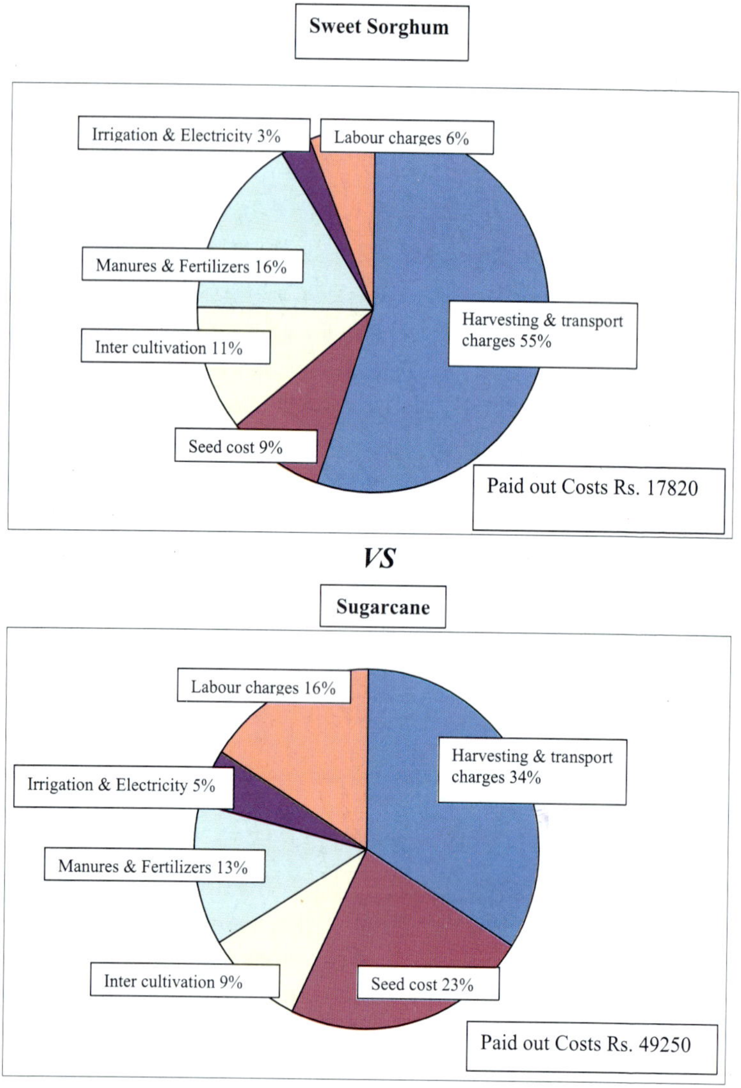

Plate No. 7. Item-wise paid-out costs of Sweet Sorghum *VS* Sugarcane cultivation in Andhra Pradesh.

Plate No. 8. Crushing of sweet sorghum canes

Plate No. 9. Checking pH of the juice

Plate No. 10. Sweet sorghum syrup

Plate No. 11. Sorghum jaggery

Plate No. 12. Ethanol industry

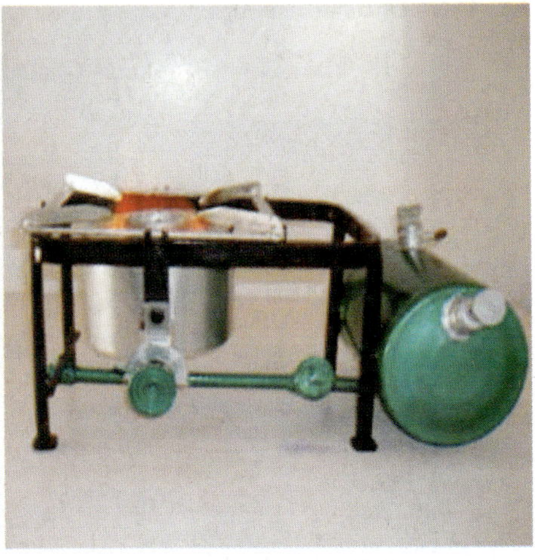

Plate No. 13. Low concentration Ethanol Stove

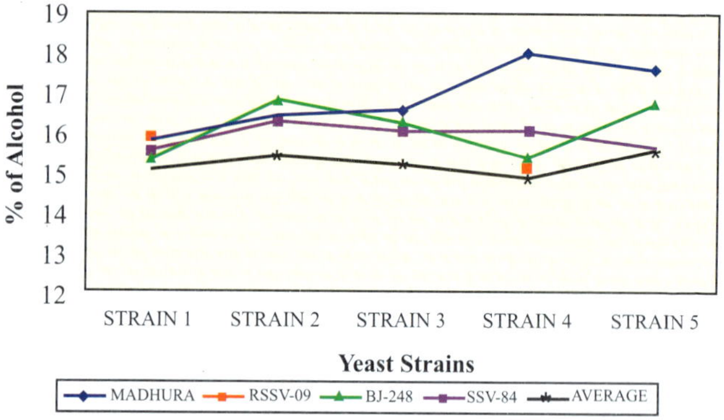

Plate No. 14 Effect of yeast strains on percentage of alcohol production from sweet sorghum juice.

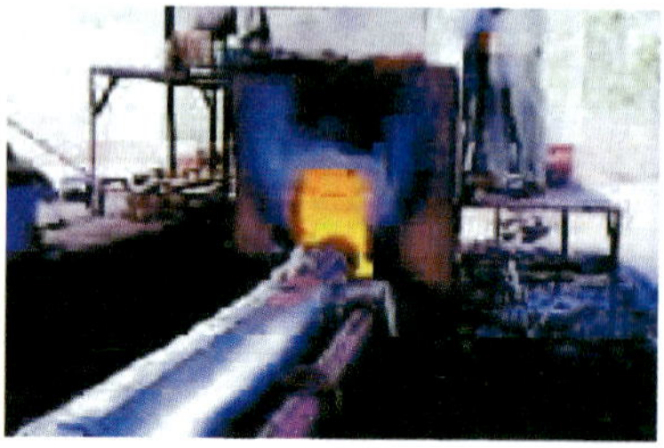

Plate No. 15 Biomass Gasifier and Furnace

Plate 16. Ethanol industry and storage containers

Index